ROSES
LOVE GARLIC

by Louise Riotte

McKenzie Steele Briggs
A.E. McKenzie Co. Ltd.

Vancouver, Edmonton, Regina
Brandon, Toronto, Montreal, Halifax

McKENZIE

Illustrations by the author.

Printed in the United States by R. R. Donnelley

Special McKenzie Edition, *November, 1987*

Library of Congress Cataloging in Publication Data

Riotte, Louise.
    Roses love garlic.

    Includes index.
    Bibliography: p.
    1. Gardening.   2. Organic gardening.   3. Companion
crops.   4. Plants, Useful.   I. Title.
SB453.R53   1983   635.9   83-1464
ISBN 0-88266-331-3

A.E. McKENZIE CO. LTD.

McKenzie is proud to be able to support new, innovative and unique methods of bringing gardening information to the marketplace. *Roses Love Garlic* is one in a series of new books McKenzie will be introducing to gardeners in the coming years. We at McKenzie are dedicated to the belief that gardening is for everyone and that gardening can be "Fun" and "Rewarding."

We have been practising this belief since the Company was founded in 1896 at Brandon by Dr. A.E. McKenzie. The seed company quickly flourished under his gardening spirit plus a high demand for quality seeds: in fact, customer requests increased to the point that in the late eighteen hundreds, McKenzie released its first seed catalogue.

The Company has grown since then and has acquired other companies to the point where it has become the largest consumer-oriented Flower and Vegetable Seed Company in Canada. The Company serves Canada from Coast to Coast and trades under the name "McKenzie-Steele-Briggs." However, recently the Company has moved to reestablish its roots and is now known as McKENZIE.

We hope you enjoy *Roses Love Garlic* and we look forward to bringing you other informative products for gardening in the future.

McKENZIE
SERVING THE HOME GARDENER
SINCE 1896

CONTENTS

INTRODUCTION

Flowers, from wildflowers to their finest hybrid culture, are beloved by people in every part of the world. Most are used as decorations, but many are also used for medicinal purposes, or to help other flowers grow or resist disease.

We also depend on flowers and flowering plants for our food. Flowering plants include almost all of our grains, fruits, and vegetables. We eat the roots of beets and carrots, the leaves of lettuce, the seeds of beans and peas, the fruits of apples and peaches, and the young stems of asparagus — all of which are blooming parts. And artichokes, broccoli, and cauliflower are undeveloped flower clusters. Even the animals that we eat — cattle, sheep, and hogs — live on flowering plants.

We have learned many different ways in which to use plants. Dandelion and elderberry blossoms are used to make wine. Cloves, the flower buds of the tropical clove tree, are used to flavor many foods. Pickled flower buds of the caper bush are used as a relish. Even the thick, fleshy petals of the Guatemalan earlflower are sometimes used to add flavor to tea and chocolate.

Many flowers are known by their fragrance. And this smell may range from pleasant to unpleasant — from the delightful fragrance of the rose to that of the pelican flower of South America, which smells like carrion.

Most flowers need soil to grow, but some can grow on tree branches taking their sustenance from the air — others float on lakes and streams. Even hot, dry deserts have many lovely blossoms. During and after the rainy season, they spring up as if by magic to bloom quickly and set seed so another generation will be there to rise again when the season is right. Just about the only places flowers do not grow are in the ice-covered parts of the Arctic and Antarctic.

Some water blossoms are so small they can be seen only under a microscope; others like the giant rafflesia, the largest flower in the world which grows wild in Malaya and Indonesia, may measure three feet across!

Flowers are generally classified by the length of time that the plant normally lives. Annuals live only one year. Biennials live for two years, blooming well only the second year. Perennials live for more than two years; they bloom the year after they are planted.

Trees and shrubs that also flower usually live for a number of years. Some plants, such as the daylily, have blossoms that live only one day. And many flowers are not only edible but also contain vitamins and minerals. Rose hips and violets are both very high in vitamin C.

With such a wide diversity, it is hard to define a "flower." The word may mean either (1) the blossom or (2) the whole plant. Botanists use the word flower to mean only the blossom of a plant. They call the whole plant—blossom, stem, leaves, and roots—a flowering plant. Any plant that produces some sort of flowers, even a tiny, colorless one, is a flowering plant. Thus, grasses, roses, lilies, apple trees, and oaks are all flowering plants.

COMPANION FLOWERS

Companion planting is not a form of magic. It is simple and practical, making use of known factors in planning a flower or vegetable garden. Companion plantings of some kinds have been practiced throughout most of agricultural history.

Early settlers from Europe found the American Indians planting corn and pumpkins together. In Holland in the 1800s a border of hemp (cannabis) was often planted around a cabbage field to keep away the white cabbage butterflies. Nature herself grows many different kinds of plants successfully as "companions." Furthermore, her plants, in most instances, grow very closely together. Instead of isolating particular kinds or varieties, she often places them shoulder to shoulder with each other. Thus they become a source of needed shade, a climbing support, or a provider of mulch and soil-conditioning food. They may even repel other plants, preventing a sturdy, too-aggressive species from completely taking over.

Legumes such as clover and alfalfa have long been used as companion crops by farmers who have grown them to add nitrogen to the soil. The nitrogen "fixed" by legumes and other plants is not immediately available to neighboring plants but is released when the legume, or a portion of it, dies and becomes incorporated in the soil. In the flower garden lupines and other plants belonging to the bean family perform the same function.

Some companion plants offer mechanical benefits; roots of large plants may break up the soil for smaller ones and make root penetration easier, especially in tight soil. Deep tap roots of dandelions and other plants bring up minerals and make them available to plants growing nearer the surface.

Large plants (hollyhocks, sunflowers) may provide shade, wind protection, or higher humidity for small plants near them. In nature, shrubs growing around the trunk of a tree may protect it from animals. At the same time the tree, because of the shade it provides, protects the shrubs from being overwhelmed by weeds.

Another key to companion planting is controlled competition. A gardener when growing a perennial flower border is engaging in one of the most complex forms of companion planting. The garden is designed not only for color, texture, height, and bloom sequence but also for controlled competition through proper spacing and varying heights. Smaller plants are protected by larger ones, but thought must also be given to the aggressive plants which will crowd out slower-growing ones if they are not kept within bounds.

We have also learned that it is unwise to plant together those plants susceptible to the same insects and diseases. Columbines, which are very attractive to red spiders, should not be planted near other flowers or tomatoes which the spider mites also find tasty.

We know, too, that certain trees exude toxic substances through their roots to inhibit germination of their own seedlings beneath them. This is their natural way to reduce or eliminate competition. On the other hand, the root exudates of dahlias are helpful against certain kinds of nematodes, and they are protective to other flowers growing nearby.

Many rock garden plants also could be considered companions because they all do well in somewhat dry sunny sites. Environmental factors make these plants companions.

Pumpkins and corn, as the Indians knew, grow well together because they are suited to the same conditions and their growing rates let them compete favorably for light, water, and nutrients. Plants that like the same growing conditions but occupy different soil strata make good companions—African marigolds and narcissus—and the marigolds also repel certain nematodes that attack the bulbs.

We find many other unusual examples of "togetherness." Indian paintbrush, a beautiful flower and a great attractant for hummingbirds, will not grow from seed (in cultivation) unless another plant is sown in the same pot with its seeds. The usual practice is to use blue gramma (*Bouteloua gracilis*).

So now, let's look at some of the more unusual flowers and plants, both cultivated and wild, which you will find beautiful, and useful.

A

ACACIA (*Acacia*) WATTLE, MIMOSA.   These tender trees and shrubs with ornamental foliage have attractive flowers in spring. They may be grown outdoors in mild climates. Some species seem to know which ants will steal their nectar; they close when ants are about, opening only when there is sufficient dew on their stems to keep the ants from climbing. The sophisticated acacia actually enlists the services of certain protective ants, rewarding them with nectar in return for protection against other insects and herbivorous mammals.

ACCENT PLANTS.   These plants are planted singly or in small groups to provide emphasis in the garden. Usually they are of distinct color or form as, for example, the Italian cypress and the Atlantic cedar as well as the handsome sugar or rock maple. To avoid monotony in a perennial border, interplant with a tall, stiff species or one with light gray foliage. These exclamation marks of the garden break up the monochrome of green. Good tall plants such as Siberian iris, lythrum, and some daylilies give the desired effect. Plants with gray foliage include silver mound artemisia, santolina, and *Dianthus plumarius.*

ADONIS. FLOWER OF ADONIS.   This flower is named for Adonis, the beloved of Venus. According to legend, the flower sprang from the blood of Adonis after he was killed by a wild boar. Use annuals and perennials for the front border and rock garden.

The flowers are yellow or red and have five to sixteen petals. Adonis plants belong to the buttercup family Ranunculaceae.

Fukuju-Kai. A new horticultural gem from Japan. One of the earliest perennials to bloom, it bears semi-double, saucer-shaped golden flowers, each with a shimmering green center. (Wayside Gardens)

AFRICAN MARIGOLD (*Tagetes erecta*).   "African" is a misnomer for these plants hail from Mexico. To defeat nematodes which attack narcissus, nurserymen often plant African marigolds as a cover crop before planting the bulbs. To achieve satisfactory control, they plant the marigolds at least three months before planting the bulb crop.

African marigolds are also planted around apple trees or nursery stock used in grafting and budding to discourage pests. Planted near roses damaged by certain nematodes, they restore vigorous growth.

AFRICAN VIOLET (*Saintpaulia*).   This is a great favorite of indoor gardeners for its beauty. To propagate, plant the leaves in slightly moistened potting soil in a margarine tub. Slip tub in a plastic bag and close. Quickly new plants will grow and form roots.

AJUGA (*Ajuga*). Ajuga is a delightful ground cover. *A. reptans* var. *metallica crispa* is especially lovely planted in small patches between the green varieties. It has deep purple foliage and deep blue flowers. Although it can grow in shade, it does best in full sun. *A. reptans* (Pink Beauty) has whorls of delicate pink flowers in May and June. *A. pryamidalis*, which is larger than the others, has deep green foliage and blue flowers with purple bracts.

AKPIK (*Rubus chamaemorus*) APPLEBERRY, CLOUDBERRY, SALMONBERRY. This perennial sends out erect shoots from a creeping rootstock. Its flowers are solitary, and its terminal leaves have three to five rounded lobes, with toothed edges. The fruit is red when unripe, amber color when mature. In the fall it is collected by Eskimos who store large quantities for winter use. The berries are eaten like strawberries with sugar and cream or used in pie or shortcake. The berries are very high in vitamin C and if frozen will retain much of their nutrient value.

ALKANET (*Anchusa*) BUGLOSS. This word is derived from the Greek *anchousa*, a cosmetic paint or stain; it may possibly have been a coloring from the blue flowers used by the ancient Greek women for eye shadow. However, a red infusion may be prepared from the roots and, as Gerard says in the *Herball*, "The Gentelwomen of France do paint their faces with these roots as it is said." The genus also provides showy biennials and perennials for borders.

ALL-AMERICA. To qualify for this label, a new seed variety must be tested at thirty sites across the country, each site representing a different climate and soil. Only those seeds that grow well and are a distinct improvement over the nearest existing variety win this citation.

In its fifty-year history, the All-America Society (AAS) has given 273 flower and 222 vegetable awards. Of these, 120 flower and 89 vegetable varieties are still in commerce.

ALLELOPATHY. Some plants release chemicals into the soil that are toxic to other plants. This phenomenon is now receiving attention as a way to control weeds. Researchers at Michigan State and Cornell hope to breed weed resistance into commercial crops in much the same way that disease resistance is instilled.

An example of allelopathy is the so-called "soft chaparral," a unique association of evergreen shrubs and trees in the semi-arid land of western North America. No more than eight feet in height, these thickets of broadleaf evergreens and stunted shrubs and trees have the amazing ability to invade grasslands and encircle themselves with dry moats of bare soil three to six feet wide.

Scientists have determined that these xerophytic (arid-climate-loving) plants release fragrant chemical compounds called *terpenes* from their leaves into the surrounding air. The soil around the shrubs absorbs the terpenes, which accumulate more rapidly during the dry season, in amounts sufficient to inhibit the germination and growth of other plants in the surrounding area. Commonly known terpenes include camphor, rosin, natural rubber, and turpentine.

ALLERGY SUFFERERS.   Linda Alpert developed an allergy-free demonstration garden outside the Tucson Medical Center's Allergic Clinic. Her garden shows that many attractive plants can be grown in desert areas without contributing to the pollen count. The garden is also attractive for another reason—the plants she chose take very little water once established.

Recommended trees include desert willow (*Chilopsis linearis*) and the lysiloma, sometimes known as the fern of the desert. Both are lacy-looking with attractive flowers and grow to about twenty-five feet high.

Cassia, jojoba, and Texas ranger (*Leucophyllum*) are among the shrubs-of-choice. Flowering ground covers include the desert primrose (*Oenothera* species) and desert verbena (*Verbena wrightii*). For flowers almost year-round, there is blackfoot daisy (*Melanpodium leucanthum*). As might be expected, there is also an assortment of cactus, agaves, and yuccas, many of which are gorgeous in blossom.

ALL-HEAL (*Valerian*) Amantilla, Setwell, Jacob's Ladder (*Polemonium*).   The common perennial border all-heal is *Polemonium caeruleum*, Jacob's ladder, or green valerian, which reaches a height of eighteen to twenty-five inches and bears beautiful blue flowers in June; the variety *album* has white flowers.

The value of the valerian lies in its roots, which are dug in spring, before the plant has begun its growth. Dry, then pulverize the roots; store the powder in an airtight container. Valerian tea is useful for many nervous disorders such as cramps, headaches, or stomach gases. The flavor is not particularly pleasant but it is sleep-inducing and tranquilizing. Use a teaspoon of root per cup and steep in boiling water. As a herbal sedative it is very calming.

Dried valerian added to bath water helps with skin troubles and has a soothing effect on the nervous system. Because of its sleep-inducing quality, a small amount is beneficial when added to herbal cushions or pillows. Pillows may also contain a mixture of dried peppermint, sage, lemon balm, or lavender with small additions of dill, marjoram, thyme, tarragon, woodruff, angelica, rosemary, lemon verbena, and red bergamot.

ALLIUM (*Allium*) GARLIC (*Allium sativum*) and roses aid each other, as do other members of the onion family, such as shallots, onions, leeks, and chives.

Flowering onions are members of the lily family Liliaceae, but there are ornamental alliums which are more decorative to plant with roses and will also provide excellent protection from aphids and other pests. *A. senescens glacum* (Wayside Gardens) is a new low-growing plant with silvery blue leaves which curl and twist in a style suggestive of Japanese art. The two-inch umbels of soft rose pink are profuse in August and September. This plant not only makes a fine protective ground cover for roses but is appropriate for edging rock walls or for the front of the border. Add to its useful qualities its hardiness and drought tolerance. Plant seeds ten to twelve inches apart.

Alliums protect roses against mildew and black spot. They also repel moles. Here are some other beneficial services of the onion family:

Onion. Repels cabbage butterflies and helps all members of cabbage family.

Chives. Good companion for fruit trees and tomatoes.

Garlic. Good against fruit tree borers.

Members of the onion family were so valued in ancient times that it is said the builders of the pyramids were paid in leeks, onions, and garlic. Onions are not only healthful, but certain members of the family such as garlic are said to be an excellent aid in preserving a youthful complexion.

ALOE VERA (*Aloaceae*). The flower of this, nature's own medicine plant, is very undistinguished, having an extremely long stem and very small flowers. The plant, with more than 200 species, is a vegetable belonging to the lily and onion families.

Cut leaves exude a juice useful as a wound dressing on a tree limb after it has been cut. The healing qualities of aloes are now widely recognized and the extracts are used in various cosmetics. It is best known for its use on burns. The juice taken internally is also healing.

ALPINE FLOWERS. These flowers know so precisely when spring is coming that they bore their way up through lingering snow banks, developing their own heat with which to melt the snow. The *Stellaria decumbens* is found at 20,130 feet in the Himalayas.

ALYSSUM (*Lobularia maritima*). The white, honey-smelling alyssums are charming with Martha Washington geraniums. Or try the Violet Queen variety with a Cecile Brunner rose. Sow outdoors in early spring. Do not cover; the seed needs light to germinate. Pot up alyssum in August for indoor bloom in November.

AMARANTH (*Amaranthus*). This is the common name of a family which includes both weeds and garden plants. The family is mostly herbs. The name comes from a Greek word meaning unfading and is appropriate because the amaranth flowers remain colored even when dried.

A member of the Amaranth family, cockscomb (*Celosia*) is very often grown as a garden flower. *Celosia cristata* bears flattish, dense heads of crimson, yellow, orange, or pink flowers and is an excellent pot plant. Another type, *C. plumosa*, grows in the form of a feather plume, and comes in scarlet, crimson, and gold. These plants add brilliant color to the garden.

AMARYLLIS. This is a genus of beautiful lilylike plants which are usually grown indoors.

AMSONIA (*A. tabernaemontana, A. salicifolia*) Willow Amsonia. This unusual and little-known perennial may be used as a specimen or toward the front of the herbaceous border. Its arching, willowy stems display narrow, glossy leaves and, during May and June, clusters of small star-shaped flowers of a strange, steel blue color. The plant grows in sun but prefers part shade, particularly in warm climates. Because it is highly resistant to wind, it grows well in the Southwest and in coastal areas. Amsonias grow slowly, are never troubled by insects or diseases, and rarely need division or staking.

Pot an amaryllis in a container only slightly larger than the bulb. Cover about one-third of the bulb with soil. For best bloom, the amaryllis should be potbound.

ANEMONE (*Anemone narcissiflora*) NARCISSUS-FLOWERED. The anemone grows in open meadows, hillsides, and alpine tundra. Its flowers appear in clusters at the top of the stem, with the white petals often being tinged with blue on the backs. The early spring growth on the upper end of the root is eaten by Aleutian Island natives and has a waxy, mealy texture and taste. Note: Some members of this family contain the alkaloid *anemonine* which causes irritation and inflammation in sheep that feed on it.

ANGELICA (*Angelica archangelica*). This decorative, broad-spreading plant is the largest garden herb. Although a biennial, it will live many years if you keep the flowers cut, but once seed develops, the plant will die. The roots and leaves have medicinal properties. The candied stems are used in confectionary, the fruits have flavoring properties, and an oil of medicinal value is derived from the roots and seeds. Dry seeds do not germinate well.

ANISE (*Pimpinella anisum*). This is a white-flowered annual belonging to the carrot family. When thoroughly dry, the seed germinates with difficulty. Therefore you will get better plants from your own fresh seed, and it will add more potent flavoring to bread, cakes, and cookies. Use the green leaves in salads as a garnish.

Anise seed germinates better, grows more vigorously, and forms better heads when sown with coriander. Anise oil attracts fish.

ANTHEMIS. The name comes from the Greek *anthemon*, a flower, and refers to the plant's profuse blooming. Use these aromatic perennials for the border or rock garden. Camomile tea is made from A. *nobilis* and a non-flowering variety of this species is sometimes used for lawns, particularly in very dry areas. It is said also to improve the health of other plants when grown close to them.

ANTHURIUM (*Anthurium*). These greenhouse plants, chiefly from tropical America, belong to the Arum family. They are grown for their brilliantly colored flower spathes in spring and summer, or their ornamental leaves. The name refers to the taillike flower in the center of the spathe and is derived from *anthos*, a flower, and *oura*, a tail. But the tail always reminds me of Pinocchio's nose!

One of the most magnificent anthuriums is A. *veitchi* which has metallic-green leaves two to four feet long.

ANTS (*Formicidae*). Ants are often a nuisance in the garden or in houses. They are repelled by pennyroyal, spearmint, southernwood, and tansy. Indoors repel them with cucumber slices.

APHIDS. Green Fly, White Fly, Black Fly, or Plant Louse.   A most familiar small pest in the garden and on houseplants, aphids come in green to match plant stems or in red or black. They are sucking insects and affect nearly all plants at one time or another. They attack houseplants in late winter when the plants' resistance is low. The lady bug beetle with her weird looking larva is always on the lookout for aphids and these larva help keep the aphids under control. There is another insect which preys upon aphids with such voracity that it is named the aphids-lion. Garlic, chives (and other alliums), coriander, anise, nasturtium, petunia, pennyroyal, spearmint, southernwood, and tansy will repel aphids.

APPLE BLOSSOM.   Louise Beebe Wilder states, "The daintiest smells of flowers are out of those plants whose leaves smell not; violets, roses, wallflowers, gilliflowers, pinks, woodbines, lime-tree blooms, and apple blossoms." The apple blossom is indeed sweetly fragrant.

Apple scab is the most troublesome disease of apples, attacking various tissues including leaf and fruit, and occurring most often in wet weather. Some varieties are more susceptible than others such as McIntosh, Delicious, and Cortland. Some people have found that chives planted around the base of apple trees, or a watering with chive tea will prevent scab. Pour boiling water over dried chives, infuse fifteen minutes, dilute with two to three parts water and use at once.

The apple tree, too, is a member of the rose family. For this reason planting members of the onion family, especially garlic, in a circle around the trunk is an aid against borers.

Onion spray is also a non-toxic fumigant, according to Beatrice Trum Hunter in her book *Gardening Without Poisons*. She states that in an experiment where a water solution of onion skin and crushed onion was sprayed on apples as they were packed in boxes, damage to fruit was significantly lower than in untreated boxes. And, within two to three weeks the onion odor had disappeared.

It is recommended that apple growers use a water spray in spring to control the codling moth. Pay particular attention to spraying loose bark to dislodge the larvae.

ARCTIC DOCK. See Dock, Arctic.

ARTHRITIS.   Eating the right fruits, vegetables, and seeds sometimes helps alleviate the pain of arthritis. These include: alfalfa, tea brewed from alfalfa seeds, asparagus, celery, cherries, collards, fennel, gooseberries, kale, lemon juice, lettuce, limes, melons, molasses, mustard greens, oranges, sage, spinach, sunflower seeds, tangerines, and watercress.

ASPARAGUS, BEACH *(Salicornia pacifica standley)*. The stem is smooth, fleshy, and jointed with opposite branches. The inconspicuous flowers are usually three sunk into the fleshy hollow of the thickened upper joints. The plant grows on sea beaches in southeastern Alaska around Prince of Wales Island and Ketchikan. It is available in summer. When young, plants may be used in salads or for pickles.

ASPARAGUS FERN *(A. plumosus)*. The plant, a member of the lily family, is slender with fernlike foliage on climbing stems. The fronds are very popular for floral arrangements. *A. sprengeri*, an ideal plant for pots, has long branched stems clothed in narrow leaves and bears small white flowers followed by small red berries. *A. medeoloides*, the smilax of the florist, has dense minute foliage.

ASPIDISTRA ELATIOR. PARLOR PALM, CAST-IRON PLANT. Gracie Fields made this plant famous in the song, "The Biggest Aspidistra in the World." During Victorian times it was probably the most popular houseplant, gradually giving way to the philodendron, dracaena, and ivy. However, it is becoming popular again, perhaps because it is the most easily managed of all houseplants and may be kept healthy and vigorous for years with a minimum of attention.

Aspidistras are shade plants with a low respiration level. Even with little sunlight the leaves can support a steady growth of all parts of the plant. Flowers come in winter, December to March, and arise at soil level. With their magenta and gold colors they are reminiscent of sea anemones or tiny exotic lilies, to which they are related. In their native forests of the Himalayan or Japanese foothills, the flowers are pollinated by a tiny snail crawling over them. As "potted captives" the plants seldom produce seeds but may be increased by root division. The types with variegated leaves of cream and green are especially attractive.

ASTER *(A. frikartii)*. The plant sends up an abundance of flowers from June to November, even after a frost or two, and deserves to be seen more often in gardens. Asters are an immense group with about 160 species native to North America. On moist, low soil or by roadsides we find bushy aster *(Boltonia steroides)*; New England aster; tradescanti aster; and willow-leaved aster; and on banks of streams and in swamps, purple-stemmed aster *(A. puniceus)*. If asters invade pastures or fields, it indicates a need for drainage.

ASTILBE *(Astilbe)*. The name is thought to be derived from the Greek word for not shining, a reference to the leaflets. Perennials are useful for border and rock gardens; the many modern cultivars are generally the most handsome, and are known as *spiraea*.

ASTROLOGICAL ASPECTS.   For centuries farmers have plowed and planted according to the signs of the zodiac. These same signs are just as effective when used for flowers. Use them in conjunction with a good gardening almanac (see Sources of Supply) which gives the correction for time changes in each part of the country.

Plant	Moon Phase (by quarter)	Sign
Annuals	1st or 2nd	Libra
Asters	1st or 2nd	Virgo
Beans	2nd	Cancer, Scorpio, Pisces, Libra or Taurus
Broccoli	1st	Cancer, Scorpio, Pisces, Libra
Bulbs	3rd	Cancer, Scorpio, Pisces
Bulbs for seed	2nd or 3rd	Cancer
Cauliflower	1st	Cancer, Scorpio, Pisces, Libra
Chrysanthemums	1st or 2nd	Virgo
Clover	1st or 2nd	Cancer, Scorpio, Pisces
Coryopsis	2nd or 3rd	Libra
Cosmos	2nd or 3rd	Libra
Crocus	1st or 2nd	Virgo
Daffodils	1st or 2nd	Libra, Virgo
Dahlias	1st or 2nd	Libra, Virgo
Deciduous trees	2nd or 3rd	Cancer, Scorpio, Pisces
Flowers for beauty	1st	Libra
for abundance	1st	Cancer, Pisces, Virgo
for sturdiness	1st	Scorpio
for hardiness	1st	Taurus
Garlic	1st or 2nd	Scorpio, Sagittarius
Gladiolas	1st or 2nd	Libra, Virgo
Golden glow	2nd or 3rd	Libra
Gourds	1st or 2nd	Cancer, Scorpio, Pisces, Libra
Hay and grasses	1st or 2nd	Cancer, Scorpio, Pisces, Libra, Taurus, Sagittarius
Honeysuckle	1st or 2nd	Scorpio, Virgo
Hops	1st or 2nd	Scorpio, Libra
Iris	1st or 2nd	Cancer, Virgo
Lilies	1st or 2nd	Cancer, Scorpio, Pisces
Moon vine	1st or 2nd	Virgo
Morning glory	1st or 2nd	Cancer, Scorpio, Pisces, Virgo
Oak trees	2nd or 3rd	Sagittarius
Pansies	1st or 2nd	Cancer, Scorpio, Pisces
Peach trees	2nd or 3rd	Taurus, Libra
Peanuts	3rd	Cancer, Scorpio, Pisces
Pear trees	2nd or 3rd	Taurus, Libra
Peas, Sweet	2nd	Cancer, Scorpio, Pisces, Libra
Peonies	1st or 2nd	Virgo
Peppers, Ornamental	2nd	Scorpio, Sagittarius
Perennials	3rd	Cancer, Pisces, Libra

Plant	Moon Phase (by quarter)	Sign
Petunias	1st or 2nd	Libra, Virgo
Plum trees	2nd or 3rd	Taurus, Libra
Poppies	1st or 2nd	Virgo
Portulacca	1st or 2nd	Virgo
Quinces	1st or 2nd	Capricorn
Roses	1st or 2nd	Cancer
Squash	2nd	Cancer, Scorpio, Pisces, Libra
Strawberries	3rd	Cancer, Scorpio, Pisces
Sunflowers	2nd, 3rd, 4th	Libra
Trumpet vines	1st or 2nd	Cancer, Scorpio, Pisces
Tubers for seed	3rd	Cancer, Scorpio, Pisces, Libra
Tulips	1st or 2nd	Libra, Virgo
Valerian	1st or 2nd	Virgo, Gemini

During the increasing light (from New Moon to Full Moon), plant annuals that produce their yield above ground. (An *annual* is a plant that completes its entire life cycle within one growing season, and has to be seeded anew each year.) During the decreasing light (from Full Moon to New Moon), plant biennials, perennials, bulb and root plants. (*Biennials* include crops that are planted one season to winter over and produce the next. *Perennials* and bulb and root plants include all plants that grow from the same root year after year.)

If you wish to save flower seed, let flower heads mature for as long as possible before gathering, but do not let the seed become so dry that it will shatter. Gather seed in a dry sign such as Aries, Leo, Sagittarius, Gemini, or Aquarius.

AURICULA. The name comes from the Latin *auricula*, an ear, and is a reference to the shape of the leaves which resemble the ear of an animal. So-called alpine auriculas are probably derived from *Primula pubescens* and what are known as florist auriculas from *Primula auricula*. Auricula itself is one of the thirty or so classes into which botanists now divide the genus *Primula*.

AZALEA (*Ericaceae*). Botanically, all azaleas are rhododendrons, but most gardeners call the smaller leaved and deciduous types azaleas. Azaleas are truly gorgeous and blossoms range in color through pink, red, white, yellow, and purple. Their long pollen stalks extend beyond the petals. A long, slender pod with hairs holds the seeds. Some of the leaves are narrow, others egg-shaped. In some azaleas, the flower has a covering of sticky hairs which keeps ants away from the sweet nectar. The plants live best in acid soil and partial shade.

The Arnold Arboretum in Boston, Massachusetts, established in 1872, was the first extensively organized effort to collect and introduce

Azaleas are one of spring-time's delights, blooming early in May and June in a wide range of colors. About 40 species grow in North America.

ornamental plant varieties from foreign countries. Of the several collectors sent out by the Arnold Arboretum, Ernest H. Wilson was the most famous. During his travels throughout the Orient, Wilson gathered one of the finest collections of the azalea varieties of Japan. Since then azaleas have been hybridized into the glorious flowers we have today.

B

BABY BLUE EYES (*Nemophila menziesii*).   This plant shares honors with catnip as a feline attractant. Louise Beebe Wilder in her book *The Fragrant Garden* says cats "will even dig the plants out of the ground." Baby blue eyes, however, deserves to be more widely planted as it makes a colorful ground cover from June to frost.

BABY'S BREATH (*Gypsophila*) CHALK PLANT.   Baby's breath is a must for dainty bouquets. In early summer, these plants bear a profusion of feathery panicles of small, starry white or pink flowers on threadlike stems, creating a delicate and beautiful veillike effect. The plant withstands cutting well and succeeds in any well-drained, not-too-heavy soil, but mix some lime into the soil before planting. G. *paniculata* (Bristol Fairy) has large panicles of pure white, double flowers. Pink Fairy produces double flowers on strong, wiry stems from June to September, and adds an airy, graceful touch when placed with larger cut flowers.

BACHELOR'S BUTTON (*Centaurea cyanus L.*) Cornflower, Blue Bonnet, Bluebottle. Actually a beautiful weed, the cornflower is of value in supplying bees with honey, even in the driest weather. On limestone soils the cornflowers are blue, on acid soil they frequently develop rose and pink flowers, sometimes both colors on the same plant. The more inclined toward red, the more acid the soil.

BALM, LEMON (*Melissa officinalis aurea*). The flowers, which are salvia-shaped, are white, small, and inconspicuous; the heart-shaped leaves are sometimes variegated green and cream. When crushed in the hand, the leaf emits a delicious odor, suggestive of lemon-scented verbena. *Melissa* is Greek for bee and bees obtain large quantities of honey from the flowers. The plant will flourish in ordinary garden soil but needs a sunny, well-drained location.

BALSAMROOT (*Balsamorhiza sagitta*). Powder of stems and leaves are somewhat toxic to pea aphids. The seeds are edible and may be roasted, ground, and mixed with flour to make a bread, according to Nelson Coon.

Balm (Melissa) *was used by ancient Arabs as an ingredient in a cordial. Many home remedies call for it to treat vertigo, migrane, lack of appetite, and indigestion.*

BAMBOO (*Gramineae*).   Bamboos are huge grasses, many with stems broader than a man and some attaining a height of thirty feet or more. Their grace is visible but not their inordinate strength. They have been used to build houses, make furniture, erect bridges and fences. And, of course, the old-time angler couldn't fish without a bamboo rod.

Once the bamboos flower and produce seed, they usually die, but this may take twenty to forty years. Even cuttings and transplants from the same generation, taken to widely separated parts of the world, cannot escape this process of unavoidable maturity and subsequent death.

Besides the tropical plant there are also hardy varieties. Although they may die down in winter, the roots stay alive and new shoots are put forth in the spring. Two of the hardiest are *Sasa palmata* (*Arundinaria palmata*) and *S. variegata*. An even hardier variety is *S.p. nebulosa*.

BANANA PEELS.   Plant, don't pitch! Tear the peels or cut with scissors into small pieces and bury around your roses. Peels provide 3.25 percent phosphorus and 41.76 percent potash. But don't overfeed; three peels per bush at a time is about right. Stockpile extras by freezing them in half-gallon ice cream containers.

BAPTISIA AUSTRALIS. Blue False Indigo.   This perennial of unique appeal makes an outstanding cornerstone in the perennial border. Its blue green leaves stay handsome all season and its nine-to-twelve-inch spikes of intense blue, pealike flowers bloom in late spring and summer. It is splendid as a companion for oriental poppies, and grows best in a lime-free soil in a sunny location.

BARKING BUSHES.   Try something very unusual like *Corylus avellana contorta* (hazel) for your next flower arrangement. Its branches are so fantastically twisted and contorted that it is almost corkscrewlike in appearance. Plant it where you can enjoy its strange silhouette against the winter snow.

BATHING BEAUTIES.   To wash off your houseplants, put them in the shower, turn on a gentle, tepid spray for a minute, then leave them a few hours so excess moisture can drip off. You'll be surprised to see how this perks up plants, especially ferns.

BAT PLANT (*Tacca chantrieri*).   According to Thompson and Morgan (see Sources of Supply), the bat plant has "the blackest flower in the world." It hails from Malaya and Burma. Some call it the Devil's Flower, and the many strange stories told about it probably originate from the malevolent way the eyes in the bloom seem to follow your every move. Sometimes its curious inflorescence looks batlike. To some it resembles an aerial jellyfish. It is indeed an awesome flower and a prize for those who want to grow something different. (Also listed in *Park's Flower Book*.)

BEAN, SCARLET RUNNER. This is the king of the ornamental beans, growing over ten feet tall with large clusters of bright scarlet flowers which blossom all summer. It is very prolific; the more pods you pick, the more the plant produces. The pods are twelve to sixteen inches long with large black and scarlet colored beans which are absolutely delicious freshly cooked. If left on the vine to mature, the beans can be made into attractive necklaces. Pierce while still green and let dry for a few days on a long hat pin. String on heavy thread with small gold beads in between.

This flowering bean is ideal for growing up the side of a porch, garage, or house as a vine for shade. The scarlet runner has unusually large leaves which maintain a lush green color all summer, and the flower attracts hummingbirds—which adds to the beauty of the scene.

Summer savory, strawberries, potatoes, beets, celeriac, and summer radishes are good companions, but do not plant members of the onion family nearby.

Scarlet runner beans are both delightfully pretty and excellent to eat. They're a fine choice if you need a vine that will grow quickly to hide an unattractive fence. Keep them away from onions and garlic.

BEE FLOWERS.   A bee is said to make three journeys in order to bring one drop of nectar to the hive; 25,000 foraging trips are said to be necessary to gather the raw material for one pound of honey.

Important honey plants are: clover, alfalfa, mustard, cabbage, buckwheat, willow herb, cotton, mesquite, goldenrod, acacia, blueberry, willow, maple, linden, locust, pear, plum, apple, and cherry.

Almost all single flowers produce a certain amount of nectar, but the following flowers produce nectar profusely and should find a place in every beekeeper's garden: wallflower, arabis, forget-me-not, borage, all members of the bellflower or campanula family, the mauve catmint (*Nepeta mussini*), heathers, heaths, honeysuckle, thyme, hollyhocks, crocus, scilla, chionodoxa, snowdrop, heliotrope, cleome, lavender, lemon balm, cornelian cherry, daphne, barberry, winter aconite, *Clematis paniculata*, mock orange, sunflower, bearberry, robinia, asclepias, hepatica, *Rhamnus frangula*, limnanthes, mignonette, phacelia, scabious, stonecrop, and the Michaelmas daisies.

In addition all the small fruits, including currants, loganberries, raspberries, strawberries, and blackberries, are valuable, and particularly the gooseberry, owing to its early flowering. And after being visited by bees these fruits will set much better crops.

Bees are not specially attracted to fragrant flowers, and their marked preference for those of blue color, that are so often scentless, bears this out.

It is said that if you go your way among bees anointed with the bitter juices of rue, the so-called "Herb o' Grace," bees will avoid you.

Willow flowers, passion flower, sunflower, and the inconspicuous blossoms of the English ivy are said to be intoxicating to bees.

BEGONIA *(Begonia).*   There are many varieties of begonias—tuberous begonias, wax or fibrous-rooted types, and those grown for their ornamental leaves such as *Begonia massoniana* (Iron Cross), as well as lesser well-known types.

All begonias grow well in pots, porch boxes, or hanging baskets. The best potting compost consists of fibrous loam, two parts; leaf mold or peat moss, one part; well-decomposed manure, half a part; and a sprinkling of sand. Add ¼ ounce of bone meal to each quart of compost. Keep the atmosphere moist and shade the plants from hot sunshine. Begonias do well planted with *Achimenes* (Gesneriads) in pots or boxes as both take the same culture and will bloom well in shade.

BELLADONNA LILY. Amaryllis Family.   The common garden amaryllis may be grown permanently outdoors in California and Florida, but in most places the large tuberous bulbs are taken up and stored during the winter. Store bulbs with caution because the alkaloids present in the bitter-tasting bulb cause trembling and vomiting if inadvertently eaten. The showy, sweet-scented flowers are typically rosy-pink and trumpet-shaped, which makes for a beautiful pot plant. Some members of the amaryllis family, such as the century plant and the Cuban and Mauritian hemp, are sources of useful fibers.

BERGENIA (*Saxifraga, Megasea*). These handsome plants, about one foot tall, have masses of decorative broad, deep green foliage, and clusters of pink flowers which appear in early spring from March to May. They are fine for the front of the border, to "face down" shrubs, as an informal ground cover, and for the rock garden.

BIBLE LEAF (*Chrysanthemum balsamita*) COSTMARY, ALECOST. Used as a bookmark, the Bible leaf provided some distraction for children to smell during long church services in colonial days. The plant will grow in some shade but will not bloom there. The flower heads are golden yellow, small, button-like, and in loose clusters.

BIOLOGICAL CONTROLS. Ladybugs and the praying mantis do good work in the flower garden, but did you know the firefly (lightning bug or lampyrid beetle) larvae benefit growers by feeding on slugs and snails? Assassin bugs feed voraciously on caterpillars, Japanese beetles, and leafhoppers. Damselflies eat aphids, leafhoppers, tree-hoppers, and small caterpillars. Flower and robberflies are excellent pollinators and the larvae eat aphids, leafhoppers, and mealy bugs.

BIRDS. Certain flowers depend upon certain insect friends to carry their pollen from blossom to blossom so they may set fertile seed; other flowers depend upon the hummingbird. Only his tongue, that runs out beyond his long, slender bill and can turn around curves, could reach the drops of nectar in the tips of the wild columbine's five inverted horns of plenty. He also seeks honey from the monarda, or bee-balm, the coral honeysuckle, the jewelweed, the cardinal flower, and many others.

Birds also play a valuable part in disseminating the seeds of many flowers. Mistletoe, for example, is spread by birds scraping their bills on the bark of trees, after they have feasted on its berries.

Birds help keep the balance of nature by trimming down the insect population. They are caretakers on the ground floor, eating grubs and beetles; they destroy grubs in the bark of trees; and others, like the purple martins, catch the flying insects.

But flowers are not always kind! The cuckoo-pint or spotted arum of Europe, a relative of our jack-in-the-pulpit, actually poisons messengers carrying her seed; the decaying flesh of the dead birds affords the most nourishing food for her seed to germinate in.

For their beauty, their song, and their ability to catch insects, birds are an asset to flower and vegetable gardens. To attract them to your garden, provide birds with food and shelter. Hedges and dense shrubs, as well as trees, provide birds with nest sites and shelter.

Birds like variety in their diet—so remember this when deciding what plants to use in your wildlife landscaping. Create a varied pattern by intermingling plant species, sizes, and shapes. Give them a choice of food sources—seeds, nuts, fruits, berries, and flower nectar. Many songbirds combine these plant foods with insects, worms, and other animal foods. Use plantings of annuals such as coreopsis, marigolds, sunflowers, and petunias for additional bird feeds. Open water of some kind is needed by most birds. This may be a conventional bird bath or a small pool with stones in the shallow edges to draw birds to drink and bathe. They will use the dry tops of the rocks for preening sites after bathing.

BIRDSEED. Feeders stocked with fruits and grains are welcomed food sources in late winter after fruits from your plantings have been depleted. A hackberry or sugar berry tree has red, edible seeds which are relished by birds in late fall and winter.

Thistle seed draws large numbers of songbirds where they have not been attracted previously. It is unappealing to squirrels and most large birds but will attract goldfinches, purple finches, pink siskins, and red polls. Tender, husked sunflower hearts also provide a wholesome food for songbirds such as chickadees, cardinals, finches, and grosbeaks.

If you wish to attract cardinals try, safflower seed. Mix it with sunflower seed at first and soon the cardinals will be won over to it, while the other birds remain indifferent. Then you'll have a feeder strictly for these colorful birds. To attract wild birds, try a traditional mix of sunflower seed, oats, millet, and yellow corn. Sunflower seed alone is the best all-around food for attracting the greatest number of desirable birds such as cardinals, chickadees, bluejays, grosbeaks, nuthatches, finches, and titmice. (*See Wild Birdseed.*)

BLEEDING HEART (*Dicentra*). It may have red, pink, or white flowers. *D. spectabilis* is the old-fashioned showy bleeding heart with long gracefully pendulous racemes covered with heart-shaped pink flowers on plants about two feet in height. This old-time favorite is still very popular. Of easy culture, these plants increase in size but do not need transplanting or dividing very often. However, since they do go dormant early in the fall, it is wise to set another plant close by as a filler; *Anemone vitifolia* is recommended for this purpose.

BLOODGOOD (*Cornus alba*) SIBIRICA. It is unbelievable for winter color. Its bright red stems provide intense contrast against evergreens or winter snow. *Euonymus alatus* (cork bark euonymus) is another prize for flower arrangements. The twigs develop pronounced corky wings which are very well defined. Plant this shrub in sun for spectacular fall foliage as well as for the color effect of the orange fruits.

BONSAI.   Bonsai are miniature trees grown in pots. The aim of bonsai culture is to develop a tiny tree that has all the elements of a large tree growing in a natural setting. Over the years, the Japanese have devised standards of shape and form which gradually became the classic bonsai styles. Many ordinary shrubs and trees take to bonsai—quince, forsythia, even a scrubby little American elm shoot.

Begin by cutting back the roots; if your plant has a tap root, cut it off to the end. Trim other roots if numerous, but not too much. Shape the top and put plant in a clay pot with a hole in the bottom for drainage. Rocks at the bottom and maybe a screen to keep out the bugs are helpful.

Screen your soil in three sizes, through quarter-, eighth-, and sixteenth-inch screens with the larger lumps at the bottom. The soil mix is a third each of sand, compost, and soil.

Set plant in the soil, water well, and in the beginning limit sunshine to mornings. Branches may be shaped by clipping, or trained on an attached wire covered with twig tape (a greenish brown color) to hide the wire. Twistems may also be used to hold branches in the unnatural position.

Plants should not become rootbound nor should the pot be too big.

Bonsai were developed in the Orient, but now their popularity has spread around the world. Some bonsai are hundreds of years old—and still tiny—but don't let this deter you from trying to grow your own shrub or tree.

The pots may be sunk in the ground to winter the plants from November 1 to March 1. If the plant is in a container that might crack, move it to a clay pot.

See U.S. Dept. of Agriculture bulletin "Growing Bonsai" No. 206, for more detailed information.

BORAGE *(Borago officinalis)* is the common name of a familiar herb whose leaves and flowers are used for flavoring claret cup and other beverages to which it imparts a cucumberlike fragrance and refreshing flavor. The blue flowers are also dried for use in potpourri. It is an annual and easily raised from seed sown in spring in ordinary garden soil.

For many centuries, borage has been used medicinally; in the preparation of various cordials and cups, it is believed to have an exhilarating effect. Pliny had a high opinion of its virtues "because it maketh a man merry and joyful." Use the young leaves in salads and the flowers as a garnish.

The plants require a sunny position, but the blue coloring of the flowers is finest when the plants are grown in poor soil.

BORERS.   Garlic planted around fruit trees will repel borers but is best done when trees are young and newly planted. Nasturtiums are also good. To foil borers in squash plants soak the seeds overnight in kerosene.

BOTANICALS.   A wide variety of plants have been used to repel moths, including oil of cade *(Juniperus oxycedrus)*, lavender, costmary, wormwood, and clove; leaves of fennel, patchouli, sweetflag, fern, bracken, and rosemary; flowers of the male breadfruit tree; black pepper; Irish moss; citron; alcoholic solution of coumarin and hemp; extract from broom seed, cinchona, lupine, tung oil, and elcampane. The wood of cedar has long been recognized as a moth repellent.

In addition to moth repellents, botanicals have been found useful against other insects which destroy cloth. These include camphor and powdered clove against carpet beetle larvae. Clothing has been treated with a soapy emulsion of anise oil or bayberry oil to ward off insects.

Other botanicals have been used to relieve domesticated animals of insects. Freshly cut pumpkin or squash leaves, a decoction of black walnut leaves soaked overnight, or an infusion of pignut leaves, rubbed on horses or cattle, will repel flies. Sometimes yellow wild indigo is placed on harnesses to keep horses free of flies. Concentrations of potato water rubbed on cattle, and clove on chickens and dogs, will repel lice, and a water solution of wormwood is used to bathe small animals and rid them of fleas. In Brazil, a tincture of cocoa leaves is considered a remedy for poultry lice, while cocoa shells, used as bedding for dogs, are credited with repelling fleas.

BOULDERS.   Tuck succulents and tiny rock plants into crevices and crannies of an ancient boulder. Rub with moss and lichen and let sun and rain be its benison. I have one and love it.

BOUNCING BET *(Saponaria officinalis)* SOAPWORT.   This showy flowering plant grows almost too readily. Its great virtue lies in its sudsing quality, the bruised leaves acting as a soap when agitated in water. The lather may be used as a shampoo.

Bouncing bet was brought to the New World more than 300 years ago for its valuable saponin qualities. It was once used extensively for washing fine silks and woolens. The carnationlike pink and white flowers cover the plant during its long blooming season.

BRACTS.   What we sometimes think of as blossoms are often petallike bracts such as the highly colored bracts of the poinsettia. The dogwood and the bunchberry have tiny purple flowers surrounded by white, petallike bracts. Bracts are leaflike structures which may form a circle beneath an inflorescence.

BROCCOLI *(Brassica oleraceae).*   The unopened broccoli flower buds are the edible part of the plant. The white cabbage butterfly lays its eggs on this plant. To protect it (and other members of the cole family), use Bacillus Thuringiensis (BT) sold under various trade names as Dipel or Thuricide. Broccoli does well with aromatic plants such as dill, celery, camomile, sage, peppermint, and rosemary.

BROMELIADS.   Did you know you can force a bromeliad to bloom by covering it for five days with a plastic bag with an apple inside? A bromeliad is any plant belonging to the pineapple family. Typical bromeliads are aechmea, billbergia, cryptanthus, nidularium, tillandsia, and vriesia.

If you're cramped for space indoors, try miniatures. Among my favorites are the little bromeliads, specifically those known as cryptanthus, or "earth-stars." The species known as *vivittatus minor*, for example, forms a rosette three to four inches across. It hugs the ground and is composed of leathery-stiff leaves, the edges of which have tiny spines. If you run your index finger lightly along one it will remind you of a cat's tongue.

The leaves are striped lengthwise in color that varies depending on age and growing conditions. Usually they are a combination of green with red or pink suffusion. The flowers, which are white and typical of cryptanthus, grow from the center of a mature plant, hence the name of the plant.

BROOMS *(Cytisus* and *Genista).*   Plant these lovely, spring-flowering shrubs in a hot corner on poor sandy or gravelly soil if you wish them to flower lavishly. If fertilized heavily, they will be barren of bloom.

Brooms come in many brilliant colors and are breathtakingly fragrant. They grow very quickly, filling in that dull patch just after the azaleas finish. The branchlets do not lose color in winter and, of course, you can make your own brooms from them!

BUCKBEAN (*Menyanthes trifoliata L.*) BOGBEAN, WATER TREFOIL.   This perennial plant has white flowers tinged with rose borne in a terminal cluster on a stalk four to twelve inches long. Rootstalks used as an emergency food must be dried, ground, then washed several times to leach out the bitter principle, and then dried again. Fernald describes the bread made from such flour as "thoroughly unpalatable but nutritious."

BUCKEYE (*Aesculus pavia*).   The flowers of the dwarf or red buckeye attract and kill Japanese beetles.

BULBS.   You can have bulbs flower in succession, from snowdrops in early spring to lilies in late summer. There are bulbs for every purpose:

   Beds and Borders.   Try hyacinths with English daisies in pink or white, pansies in selected colors, or forget-me-nots for color combinations. Tulips give a medley of color in a good-sized tulip bed edged with pansies.

   Rock Gardens. Small bulbs bring color and early season interest. Snowdrops lead off; then come the crocus species in white, yellow, and lilac shades; followed closely by little *Iris reticulata* in very dark purple

The buckbean, found in marshes and bogs, is used in many ways. Its roots can be dried and ground, then used in an emergency for flour. In Europe the roots have been used to replace hops in making beer. Extreme bitterness is a problem when using this root for food.

with orange veinings. Other suggestions include grape hyacinth (Muscari), glory-of-the-snow (Chionodoxa), spring glory (Scilla), small narcissus (*N. minimus*), hoop-petticoat daffodil (*N. bulbocodium*), and angel's tears (*N. triandrus*), to name but a few possibilities that do well together.

Forcing Bulbs for Winter Bloom. Narcissus, such as paper-whites and Chinese sacred lily, are the easist and earliest to bloom. Set them in bowls of pebbles and water in September and they'll bloom for Thanksgiving. Later starts will prolong the season. Hyacinths, white Roman, and the miniatures in yellow, pink, and blue go well in bowls of vermiculite or the special bulb fiber obtainable from dealers. Daffodils and tulips are mostly grown in standard pots or bulb pans in a good soil mixture. Use pre-cooled daffodil bulbs or cover pots outside until after a heavy freeze before trying to force them.

After blooming, spring bulbs should have the dying flowers cut off, though not the whole stem, so that the plant's strength is not wasted in seed production. Work the soil lightly between the bulbs with a hand fork or hoe before planting with annual bedding plants for summer display.

Naturalizing Bulbs. Some bulbs grow and flower for many years under natural or "wild" conditions; their only need is good soil. Snowdrops often grow and flower on the north side of a slope for generations. Grape hyacinths grow and spread. So will Siberian squills and glories-of-the-snow. Crocuses bring color and interest to bare ground. Trumpet or large cupped daffodils look enchanting on a grassy slope. Plant poet's narcissus (*Narcissus poeticus*) near water. Spanish bluebells (*Scilla hispanica*) in delicate shades of white, pink, and blue enliven the somber green of ferns in a moist, shady spot. Daffodils naturalize readily although they will not increase as rapidly in grass as under cultivation.

BUMBLEBEE PLANT (*Pedicularis lanata*) WOOLY LOUSEWORT. This perennial flowers in spikes that are pink to rose although occasionally they may be white. The entire plant except the lower leaves is densely gray and woolly. It is common on the tundras of the high mountains and in the Bering Sea district. Flowers are collected in June by the natives around Cape Prince of Wales and Shishmaref. Water is added and the flower allowed to ferment. The root is also edible, and may be gathered in the fall and prepared by boiling or roasting.

BURBANK, LUTHER (1849–1926). Burbank was an American plant breeder and horticulturist who developed many new trees, fruits, flowers, vegetables, grains, and grasses. Among the plants he developed are the Burbank potato, the giant Shasta daisy, the spineless cactus, and the white blackberry. He also improved many plants and trees already known.

One of the methods he employed was the selection process, continued through many generations until a superior plant, with respect to a single character or group of characters, became isolated. He employed this method in the development of the "stoneless" plum.

First, Burbank selected from a large collection of plums one which had a thin "pit" or stone. He permitted this tree to carry on pollination and set fruit. Then he chose from this tree a few plums that had the thinnest stones. He planted the seeds from these, and when the resulting trees fruited, he again selected the fruits with the thinnest stones. These were planted and when the resulting trees matured, Burbank again selected the fruit with the thinnest "pits." After several selections of this type, he produced a plum with an extremely thin stone; this was marketed as a "stoneless" plum. In using this method, only the desirable variations are retained. Burbank also used this method in developing the Shasta daisy. There is really no magic about the selection process; anyone with time and patience can use it.

Burbank's first discovery was the potato which bears his name. As a young gardener in Massachusetts, he planted the contents of a seed pod from the rarely blooming Early Rose potato. The pod produced twenty-three widely varying seedlings. Burbank sold the best plant to a dealer, who named it the Burbank.

BUTTERCUP, PALLAS (*Ranunculus pallasii schlecht*) Eskimo Kapugachat. This pretty plant is found growing in shallow edges of tundra lakes and ponds. The natives of the lower Kuskokwim Valley use the young tender succulent shoots which are available in spring and autumn. These young shoots grow in the saturated sphagnum moss at edges of tundra lakes and ponds. Cook before eating to drive off the anemenol contained in the plant which is poisonous. (Native of Alaska.)

BUTTERCUP FAMILY (*Ranunculus*). Root secretions from these plants poison soil for clover by retarding growth of nitrogen bacteria. Clover will disappear in a meadow where buttercups are increasing. Delphinium, peonies, monkshood, columbine, and double buttercup belong to the same family. Unless the soil is very rich, other plants will not grow well in their vicinity. This is a strong, vital family, but they grow only for themselves.

BUXUS (*Buxus*) Boxwood. Because of their handsome appearance and beautiful foliage, the common box and the edging box are greatly valued as garden ornamentals. Most will thrive in any good garden soil, but are particularly useful for planting on limestone ground. Many are grown for topiary work either to stand out individually or for hedges and dwarf borders for garden beds and paths.

Buxus Green Velvet is a new hybrid boxwood, a cross between English and Korean box and is the result of a breeding program in Ontario, Canada. The foliage, a lush bluish green overlaid with a velvety warm texture, is of unique coloring and will withstand even the harshest winter. Use for hedges, foundation planting, or specimen plants. (Wayside.)

C

CACAO.   The source of chocolate is the seeds or "bean" of the cacao tree. Native to tropical America, the trees have been cultivated for more than 4,000 years. After the beans are dried they are shipped to chocolate factories, cleaned, roasted, and ground into a pastelike substance called chocolate liquor. Pressing out the fat from this produces dry cocoa.

Researchers recently discovered that cocoa contains sizeable amounts of phenylethylanine, a substance produced by human brain cells during emotional episodes.

CALIFORNIA BUCKEYE *(Aesculus californica).*   Flours that are made with meathulls of the nuts are toxic to larvae and adults of Mexican bean beetles; also certain parts of this plant are toxic to humans.

CAMOMILE *(Anthemis nobilis).*   This plant of the aster family has strong scented foliage and flower heads which contain a bitter medicine principle. It is used as an antispasmodic, stomachic, and perspiration ingredient for breaking fevers.

Camomile in small quantities increases the essential oil in the peppermint plant, but as the proportion of the camomile plants increase, the amount of peppermint oil diminishes—too much of a good thing!

Camomile tea is effective against a number of plant diseases, especially in young plants. It can be used to control damping-off in greenhouses and cold frames. Make tea by soaking dried blossoms for a day or two in cold water.

Rayless camomile *(M. matricarioides).* The powdered heads are fairly toxic to diamondback moths.

Scentless false camomile *(M. indora* or *M. camomile).* The flower heads are said to be as effective as commercial pyrethrum in controlling face flies.

CANADIAN FLEABANE *(Aesculus pavia).*   Flowers from this shrub attract and kill Japanese beetles.

CANDYTUFT *(Iberis)*. This plant is practically foolproof even for "purple thumbers," and does well in any soil in sun or light shade. To extend the blooming period, shear off the spent blossoms. Use this evergreen subshrub for edging borders and garden walks, in the rockery, or for mass plantings. Little Gem and Snowflake have white flowers; *I. jucunda* is noted for its dense pink flowers and blue green foliage.

CANNA *(Canna)* INDIAN SHOT. These tender herbaceous perennial plants from South America and the West Indies have unbranched, stately stems springing from a fleshy rootstock. During the summer they have large, ornamental foliage and brilliant, showy gladioluslike flowers in dense terminal clusters.

The leaves and stems of cannas have insecticidal properties. They are particularly useful as a greenhouse fumigant to control aphids, ants, and mites. Place newspapers in the bottom of a pail with a metal grating on top of them. Above the grating put a layer of straw to act as a buffer; then place leaves and stems on top. Light the newspapers; the leaves will smolder rather than burn, producing thick smoke. Close the greenhouse door tightly and smoke the leaves for about thirty minutes.

CARNATION *(Dianthus)*. These charming plants of cheerful colors and spicy fragrance have attractive, often bluish green foliage. They are fine for borders, edging, and rock gardens. The hardy border carnation, *D. caryophyllus*, blooms almost perpetually. Grow it in well-drained soil with lime added and protect it in winter with a dry mulch. The English Giants produce red, rose, pink, white, and speckled blossoms in June and July which are distinguished by an especially sweet fragrance. Cut back the plants after blooming.

Carnations are very popular for Mother's Day.

CARNIVOROUS PLANTS. INSECTIVOROUS PLANTS. These plants trap insects for food. They usually live in moist places where they get little or no nitrogen from the soil. They must obtain it from the decaying bodies of the insects they trap. For this they have special organs and glands that give off a digestive fluid to help them make use of their food. Some of these plants have flowers that are colored or scented like decaying meat which helps them attract insects. Pitcher plants have tube-shaped leaves that hold rainwater in which the insects drown; other carnivorous plants have rosettes of leaves with sticky hairs such as those borne by the sundews. Some plants such as the bladderworts grow in water. The Venus's-flytrap make interesting house plants; in the absence of insect prey they are usually fed tiny pieces of meat, generally hamburger.

If you grow carnivorous plants as houseplants, water them with distilled or soft water to avoid toxic salt build-up.

Venus's-flytrap is a meat-eating plant. Its leaves consist of two hinged lobes. When an insect is attracted to a leaf, the lobes snap shut, and the insect is digested by the plant.

CARPETWEED (*Mollugo verticillata*). This is another low-growing weed which forms mats in gardens and on paths. It thrives well on lighter, sandy soils, but will not resist hoeing and cultivation.

Ben Charles Harris in his book *Eat the Weeds* states that carpetweed is good steamed or boiled, or used as a pot herb.

CARRION CACTUS (*Stapelia*). Not only does this cactus look and smell like carrion but it attracts carrion flies from a long distance. Blue-bottle flies also seek it out and have been known to lay their eggs on the leathery petals as they do on another foul-smelling plant, *Amorphophallus vivieri*.

CARROT, WILD (*Daucus carota*) Queen Anne's Lace, Bird's Nest. The tiny white flowers are exquisite used in flower arrangements with larger, coarser flowers. Allowed to go to seed it will become a pest but the seeds serve as seasoners for soups, stews, and baked fish; fresh or dried, the seeds make a fair substitute for anise or caraway seeds.

The herb is the wild prototype of our table carrot. If it is found in rich soil, the root is sweet and most palatable; located in sandy, hard soil, it is small and hard. Serve steamed or cooked in a little water, or cut into inch lengths and add to a soup or stew.

CARROTS. Carrots are a mild stimulant for women and a relaxant for men. However, they are mainly important as a source of vitamin A. As a matter of fact they are the only vegetable with visible vitamins; their characteristic color is the material from which the body extracts vitamin A. It promotes good vision, good digestion, and immunity to colds and infections. It also prevents sensitivity to bright lights; dry, lusterless skin and hair; acne; and warts.

A 5½-inch carrot contains 6,000 units of vitamin A. Carrots also provide salt and vitamin B-1 and, if you eat or cook the tops, iron and vitamin C. And there are only twenty calories per carrot stick.

CARVER, GEORGE WASHINGTON. Jerry Baker didn't originate the idea of talking to plants; George Washington Carver did and went him one better. Farmers' wives brought Carver their ailing houseplants, he cared for them tenderly and sang to them. During the day he would take them out to "play in the sun." When he returned them to their owners he would tell them gently, "All flowers talk to me and so do hundreds of little living things in the woods. I learn what I know by watching and loving everything."

Not only did Carver find many uses for the peanut but from the leaves, roots, stems, flowers, and fruits of many plants he coaxed 536 separate dyes which could be used to color wood, cotton, linen, silk, and even leather. Forty-nine of them were produced from the scuppernong grape alone!

CATCHFLY. This group of plants often infests dry meadows, clover, and alfalfa fields. It may become a pest if not controlled by early cutting. The sleepy catchfly, forked or hairy catchfly, and night-flowering catchfly are all annuals and propagate by seed. The sleepy catchfly opens its flower only to bright sunshine. The name "catchfly" is derived from a gluelike substance on the stem, which does catch flies.

CATS. Although cats sometimes do catch birds, their hunting instinct also leads them to keep the garden clear of snakes, mice, rats, grasshoppers, and tarantulas.

CAULIFLOWER (Brassicaceae). The undeveloped flower is the edible part. Dust with Bacillus Thuringiensis to control the white cabbage butterfly.

CELANDINE, GREAT (Chelidonium ma jus). This plant grows here and there in barnyards, pastures, and roadsides. It contains a yellow, slightly poisonous juice, which was once used against warts (hence the popular names wortweed, killwort, and devil's milk). There is a caustic quality in the juice (even more in the roots than the upper parts) which could be of medicinal value for skin conditions. It is also recommended that the freshly oozing juice be applied daily to corns until cured, or for use on sores of horses. Cultivation and early cutting before it goes to seed will keep celandine from spreading.

Grow catnip or baby blue eyes as a treat for your cat. And if the cat claws furniture, break this habit by rubbing crushed rue on the furniture, and the cat will leave it alone.

CERATOSTIGMA *(Plumbago larpentae).*   This plant is truly one of the most rewarding of all perennials. Growing only six to eight inches high, it forms mats twelve to eighteen inches across, solidly covered with clusters of intense peacock blue flowers in late summer and fall. Simultaneously the interesting leathery foliage turns deep mahogany red. Use the plant for ground cover or for underplanting shrubs. It will grow virtually anywhere, good soil or bad, well-drained or poorly drained, heat or cold, sun or shade!

CERCIS. Red Bud, Judas Tree.   Red bud grows wild in Oklahoma and Texas. It is unbelievably beautiful in early spring when every branch and twig is covered with bright violet red flowers. Transplant small specimens in early spring. Shrubby in growth, it seldom attains a height of more than twelve to fifteen feet, and may be grown as a large shrub or small tree.

CHEROKEE ROSE *(R. camelia).*   This charming decoration of our southern states originally came from China, but is now widely and permanently naturalized. The foliage is evergreen and shining and the immense white single blooms have the rich fragrance of the gardenia.

CHICKWEED *(Cerastium arvense).*   The field-mouse-ear chickweed is a beautiful flowering native plant used in gardens. It derives its name "mouse-ear" from the shape of its leaves. The blossom is large, white, and star-shaped. Because of its creeping rootstock (every joint can produce a new plant), it can become troublesome in pastures. Plow and cultivate to control it. Beatrice Trum Hunter in her book, *Gardening Without Poisons*, says that "rye overpowers chickweed." Chickweed likes roadsides, sunny hills, and even grows way up on high mountains.

Common mouse-ear chickweed (*Cerastium vulgatum*), which propagates only by seed, has smaller leaves and blossoms than the field-mouse-ear. It likes to grow in fields and on roadsides. Early cultivation in grain fields will lift up the shallow roots and so eradicate this perennial.

CHINABERRY (*Melia azedarach*).   This shade tree repels grasshoppers and locusts. Make a repellent tea with either live or dried leaves. The powdered fruit is somewhat toxic to European corn borer larvae.

CHINESE LANTERN (*Physalis*).   Chinese lantern is often grown for its large, showy calyxes, which are attractive as winter decorations. The white flowers are followed by dense clusters of bright orange scarlet, lanternlike husks enclosing scarlet berries. Cut the stems in autumn and dry to preserve. Chinese lantern is an aggressive plant which will rapidly "take over," so grow it in a waste spot to prevent crowding other flowers.

CHINESE WINGNUT (*Pierocarya stenoptera*).   The powdered leaves of this ornamental tree are slightly toxic to Mexican bean beetle larvae.

CHIVES, WILD (*Allium schoenoprasum L.*).   This edible perennial reproduces by small bulbs and seeds. The narrow, hollow leaves grow to a foot long. The flowers cluster in a rose purple umbel or head, at the top of the stem. Wild chives are found in interior Alaska and southward. Use cut leaves as a substitute for salad onions.

CHOCOLATE FLOWER (*Berlandiera lyrata*).   This branched perennial has pale yellow, daisylike flowers which smell like chocolate. The underside of the "petals" have brown veins. The large green bracts below the flower are attractive for dried arrangements.

CHRISTMAS ROSE (*Helleborus niger*) CROWFOOT FAMILY.   This perennial herb is often cultivated in gardens for its mid-winter bloom. Because it blooms in early spring in some regions, it is often incorrectly called the Lenten rose. The white or pink white flowers, about two inches across, become purplish with age. The thick but fibrous rootstalk which is blackish brown, yields drugs for commercial use. However, the rootstalk is violently poisonous if eaten and, as a warning, emits an unpleasant odor when cut or broken. It has a bitter, slightly acrid taste. The poisonous leaves may cause dermatitis on contact.

The Christmas rose is popular because it flowers when little else is in bloom.

CHRYSANTHEMUM (*Compositae*). The name is derived from the Greek *chrysos*, gold, and *anthemon*, flower. There are 100 species of both annuals and perennials, some of which are known as *pyrethrum*. Chrysanthemums are protective to strawberries. Chrysanthemums themselves may be protected by an all-purpose spray made from three hot peppers, four large onions, and one whole bulb (large) of garlic, ground together. Cover with water and allow to stand overnight. The following day strain mixture through a fine sieve. Add enough water to make a gallon of spray. Spray may also be used on roses and azaleas. Bury mash around roots of plants.

CINNAMON VINE (*Dioscorea batatas*) CHINESE YAM, CHINESE POTATO. This quick-growing vine will hide an unsightly area as it ranges up to thirty feet in a single season. In July and August it puts out profuse white, cinnamon-scented flowers borne in loose clusters. The roots are large tubers, potatolike in flavor and considered edible in the tropics. The leaves are shiny and quite attractive. Its flowers are borne on the axils where little tubers about the size of a pea also appear. These tubers, sown like seeds, will produce a full-sized vine the second year (see Proliferation). Cinnamon vine likes the sun but is not at all capricious as to soil. For quick growth, start the small tubers indoors in pots.

CIRCUS IN TOWN? Check around; you may find exotic animal manure available for the taking. If fresh, it's fine for the compost heap. Let it age a bit before using around plants, or dig it in freely between rows. Bengal tiger and lion manure may be just what you need to make your garden a roaring success!

CLEMATIS VITALBA. This vine along with hemp, coumarin, *Atractylia ovata*, and a decoction of persicary has been used to ward off weevils in grain. Protect stored grain with an oil coating of coconut, lemon, or mohwa. Large-flowered hybrid clematis is one of the loveliest vines known and blossoms abundantly in many colors.

CLINGING VINES. Vines growing on masonry walls or on trellises on wooden walls add insulation. In summer they lower indoor temperatures by protecting the outside walls from the sun's direct rays. Choose deciduous leafy vines (they will drop their foliage in winter to let the sun warm the dwelling), and plant them on the southern and western walls.

In cold weather, evergreen vines on the north surfaces will fend off the wind and keep the inside warmer. Try Boston ivy or Virginia creeper for summer insulation and an evergreen species such as English ivy for cold weather protection. Clinging vines are not recommended for wooden walls because their stems and tendrils hold moisture, causing the wood to deteriorate. Achieve the same insulating effect with trained twining vines such as wisteria or climbing roses on trellises.

CLONE. Recently the word "clone" has been appearing often in the news. In actual fact plants have been cloned for thousands of years. The word itself comes from Greek and means a twig or slip. Cloning plants is very easy because many plant cells have the power of regeneration.

With onions, for instance, I put cloning to work. When my onions sprouted, softening gradually in the process, I used to throw them out. No more. I plant these onions in the fall, sometimes in the garden, sometimes in my rose bed. From one onion a cluster of five or six will grow, delicate in flavor, and giving me early table onions the following spring.

The first year I pulled them all but in succeeding years I left several clumps as an experiment. In time they grew large. I harvested them at maturity with the rest of my crop. They kept very well but in time sprouted again and were replanted. A good onion—like an old soldier—never really dies! It doesn't even fade away.

CLOTHESPINS. Use snap or spring clothespins to train fuchsia and small vines in pots and baskets. Catch the stem through the indentations and clip the jaws to other stems, pot rims, or stakes. Also use to hang along stems as weights if you are working toward a trailing plant. Remove after a few weeks and the stem will stay in place.

In the garden use snap clothespins—the type with a spring and two grooves—to train grape vines on the trellis. The innermost groove should fit over the wire and the outer one around the stem. As the vine grows adjust the pins. Remove at pruning time when they are no longer needed. This idea works well with espaliers or with anything you need to train to wires.

CLOVE *(Myrtaceae)*. This is the name given to the flower buds of a tropical tree. The tree grows wild in the Moluccas, or Spice Islands, and in Sumatra, Jamaica, the West Indies, and Brazil. The tree's purplish flowers grow on jointed stalks and are picked before they open. Reddish when first picked, they turn dark brown when dried. The dried buds are used as spices. They have a fragrant odor and a warm, sharp taste. Maybe you know them best decorating a fine ham.

CLOVE PINKS. This is the gillyflower of medieval Europe, popular for its variety of color and sweet, spicy fragrance. Before oriental spices became available to everyone, the flowers of clove pinks were used with foods and for flavoring wine and vinegar. These lovely old-fashioned pinks bloom from early summer until late fall and keep the air delightfully perfumed—and the plants are winter hardy.

CLOVER *(Leguminosae)*. For cover crops, there are many leguminous plants among the clovers, including the common ones of red or sweet clover. There are also bur, crimson, Egyptian, and Persian or Wood's clover, but red clover is the most important member of the family. The flowers of red clover will not be fertilized unless a bee pollinates them. When red clover was first planted in Australia, there were no bumble-bees to carry the pollen; not until they were introduced did the red clover produce seed. Crimson clover is much used for soil improvement. Its flowers are often red but may be white or yellow. Clover honey is very delicious and one of the best known flavors.

Clover fixes the nitrogen in the air by means of bacteria growing on its roots. When the clover is plowed under, the nitrogen enriches the soil.

Red clover herb tea is especially good for canaries. To make: Steep two teaspoons in ½ cup hot water allowing to stand fifteen minutes. Put a few drops a day in the canary's drinking water.

COFFEE BUSH (*Coffea*).   This dwarf variety grows just three feet tall. It has shiny dark green foliage, white starlike fragrant flowers, followed by real coffee berries which are harvested when they are scarlet. Then remove the fleshy outer pulp and dry the "beans." Freshly roasted and ground they make real coffee. (Thompson & Morgan)

Roasted chicory not only cuts down the caffeine content, but also adds body and smoothness, which many coffee drinkers prefer.

You've seen chicory growing along roasides. It's also grown commercially. A common use of the roots is to dry them, grind them, then use them to make a coffee-like drink.

COLD FRAME.   A cold frame on wheels allows the gardener to place it in sun or shade as the season requires. Make the sides of the frame out of ¼-inch plexiglass. Attach with screws to 2×4s supporting a plywood bottom. The top is a plexiglass sheet with 2×2s attached underneath to hold it up and allow air to pass through. When turned over it is sealed shut, keeping out animals and insects. Use old tricycle or bicycle wheels for the frame. And just think, if there's a late freeze coming, you can wheel this cold frame into the garage at night to keep it warm!

COLD RESISTANT ANNUALS.   This term describes the first annuals to plant in the spring and those that will survive nippy autumn days. These annuals are also used for winter plantings along the Pacific Coast and in the Gulf states.

Pansies head the list. They are hardy and available in early spring in much of the country; some are even ready for late fall planting in the mildest climates. If you need fragrant annuals for your garden, try stocks. These delicious annuals are fine for cutting and last well indoors. Snapdragons resist the cold and begin to provide their colorful spikes early in the gardening season. For strong yellows or oranges, plant calendulas. The newest hybrids are compact, floriferous, and very hardy.

Larkspur or delphinium, annual poppies, centaurea, dusty miller, annual phlox or asters, primulas, cineraria, dianthus, or carnations all will perform enthusiastically for you in the early spring and will continue through several fall frosts.

COLEUS (*Coleus*).   These are superb, colorful foliage plants for shady spots in the garden—and they are fine houseplants as well. The luxuriant foliage displays shades of red, green, crimson, yellow, white, pink, and combinations thereof. As exotic in color and form as they are, coleus are probably due for even bigger changes as they react strongly to radioactivity, and many new forms have been seen at the atomic radiation center outside Knoxville.

COLTSFOOT (*Petasties frigidus*).   The tawny-colored flowers, appearing before the leaves expand, are not showy. The leaves are palmate or somewhat triangular in shape, green, shiny above and felty beneath, and may become extremely large. Widespread, the plant is usually found on tundra. The young leaves are collected by Eskimos and mixed with other greens. Mature leaves are sometimes used to cover berries and other greens stored in kegs for winter use. Fernald reports that in Eurasia the young stalks and flower heads are cooked and eaten. They are considered tasty. Euell Gibbons states that coltsfoot is also made into cough drops, cough syrup, and various teas.

*The columbine, a hardy peren-
nial, has attractive blooms in
many colors. Be careful where
you plant it. People and hummingbirds admire it—but so do
red spiders.*

COLUMBINE (*Aquilegia*).   These hardy, perennial plants bear spurred, beautifully colored flowers from May to July. They grow wild in North America, Siberia, and other north temperate countries and belong to the buttercup family, Ranunculaceae. The word *Aquilegia* is derived from *aquila*, an eagle, an indication of the spurlike petals.

Columbines will thrive in ordinary garden soil and are easily raised from seed. A sunny or partially shaded location suits them. But keep them to themselves—they do not companion well with other plants and are also very attractive to the red spider. However, the hummingbird finds their red and yellow bells irresistible.

COMFREY (*Symphytum officinalis*) KNITBONE.   This plant aids the knitting of fractured and broken bones and has been used for this purpose for centuries. It is good also for lung disorders, internal ulcers, external ruptures, burns, and splinters. The leaves make a poultice for bruises, swellings, and sprains. Flowers are pale blue pink, bell-form, and borne in drooping clusters.

Grow comfrey in the garden. Besides being medicinal it is beneficial to other plants—a sort of "plant doctor." Because it is deep-rooting, comfrey does not rob minerals in the surface soil from other nearby plants. It keeps the surrounding soil rich and moist and gives protective shade and shelter with its large, rough leaves. Comfrey is rich in vitamins A and C.

COMMONEST FLOWERING PLANT. The most widely distributed plant in the world is *Cynodon dactylon*, a toothed grass found as far apart as Canada, Argentina, New Zealand, Japan, and South Africa.

COMPASS PLANT *(Silphium laciniatum)* ROSINWOOD, PILOTWEED. This ancient Greek name refers to the resinous juice of these plants. Although a genus of fifteen species of perennials in North America, only two are likely to be seen in cultivation. Frontiersmen and hunters in the prairies of the Mississippi Valley were the first to notice this plant whose leaves, pointing north and south, accurately indicate the points of the compass. The plant bears yellow flowers which look much like sunflowers. It is a coarse plant and sometimes grows to be 10 feet tall. The leaves are about 1½ feet long, and are cut into several lobes. The petioles, or leaf stalks, bend so that the leaves, by pointing in a north-south direction, escape the strong midday sun, but get the full early morning and late afternoon light.

COMPOSITE or COMPOSITAE. This family is the largest and most highly developed of flowering plants. It consists of more than 20,000 species of herbs, trees, vines, and shrubs. Composite plants have efficient methods of reproduction. They produce many seeds and have good methods of scattering them. Some, such as calendula, camomile, wormwood, tansy, and arnica, are used to make drugs. Chrysanthemums, asters, and dahlias are grown for their beauty; others are weeds and wildflowers such as ragweeds, goldenrod, sagebrush, thistles, and burdock.

COMPOSTING FOR SOIL ENRICHMENT. The importance of abundant soil humus is to create a favorable environment for root growth. A soil in good tilth has a granular structure which readily permits rain and air to penetrate and also retains moisture for long periods to permit maximum growth. It's good garden practice to use compost in combination with mulching, and you can start a compost pile at any time of the year.

Build your pile in a pit, trench, free-standing stake-and-chicken-wire form, or any large container such as a plastic trash barrel. Build the pile with layers of organic matter in such a way that they will decompose easily.

Use grass, weeds, leaves, stalks, branches, and wood chips. Use any organic matter but avoid garbage unless it's something like carrot tops, lettuce, or coffee grounds; garbage attracts rodents. A shredder to grind up the material will appreciably shorten the length of time for the compost to ripen.

Put down a six-inch layer of organic matter, sprinkle it with finely ground agricultural lime. Add a one-inch layer of dirt if you have it, to introduce the micro-organisms needed for speedy decomposition.

Layer these materials up to about four feet high and keep the entire pile moist. Make the top of the pile slightly concave so that moisture will seep into the pile. Turn the pile to speed decomposition of organic material; this occurs when air is reintroduced into the compost.

Is all this worth it? Yes. A four-inch layer of compost worked into the soil to a depth of six inches will almost guarantee gardening success with both flowers and vegetables.

CORIANDER *(Coriandrum sativum).*   Grown for its savory seed, coriander is not suitable for the flower garden because of the evil smell of the foliage and fresh seed. However, the foliage is delicate and lacy and the rosy white flower umbels are beautiful. The ripe seeds are fragrant and the odor increases as they dry. Use them in cooking.

CORN, ORNAMENTAL.   The so-called Calico or Rainbow corn is often used in decorations around Thanksgiving time. The multicolor hybrids are a blend of many new varieties. Red Strawberry popcorn has tiny 1½-inch, strawberry-shaped ears crowded with small deep crimson kernels. Not only is this an attractive decorative variety, but a good popping corn as well.

CORN AND SUNFLOWERS.   A reader in *Troy-Bilt Owner News* recently reported an interesting experiment: "If you want to grow popcorn and sweet corn on the same lot, plant a row or two of sunflowers between them and you will have no trouble with cross-pollination." Since corn is wind pollinated, the sunflowers serve as a buffer zone between the rows.

To discourage corn earworm and black beetles, apply several drops of mineral oil to the silks at the tip of each ear as soon as they begin to dry and turn brown. Do not do this earlier or pollination will not take place and kernels will be undeveloped. Apply about three times at weekly intervals as all ears do not develop at the same time. I use an oil can for application.

Check your seed catalogs for corn with tight husks. One of my favorites is Faribo Golden Honey (Farmer Seed & Nursery), a tasty corn with long, tight husks which protect the ear tips.

Corn smut bolls are prevalent during spells of warm humid weather. If you find them in your corn patch, they should be promptly picked off and burned so they will not spread.

CORN COCKLE or PURPLE COCKLE *(Agrostemma githago).*   This is a very nasty weed in the grainfield, particularly in winter grain. An annual, it propagates by seed. It mixes with the grain in harvesting and threshing, and, due to its poisoning effect, it spoils the grain for feeding and flour. The black seeds are somewhat similar in appearance to caraway and are particularly poisonous to sheep, pigs, rabbits, geese, ducks, and poultry. The chaff from threshing should be checked for its presence before being fed to animals.

The rather large flowers have five reddish purple petals, slightly mottled on the edge with dark spots inside. If grainfields become infested with purple cockle, hand pulling may be necessary.

CORNELIAN CHERRY (*Cornus mas*). This shrub will cheer you up when the delightful fluffs of yellow bloom dot every leafless branch in February. They are followed by green foliage and, in turn, by scarlet fruits attractive to birds and great in jellies and preserves. The purple red, fall foliage makes this shrub of year-round interest. Specimens can be pruned to produce alluring stem and bark patterns. In spring cornelian cherry is also excellent for forcing.

COSMETICS. Queen Elizabeth of Hungary, an eternally youthful beauty, attributed her marvelous looks to an herb tonic which became known, in her honor, as Hungary Water. Here is how it was made:

HUNGARY WATER

12 ounces rosemary	1 ounce balm
1 ounce lemon peel	1 pint rose water
1 ounce orange peel	1 pint spirits of
1 ounce mint	wine

Mix together and let stand for several weeks. Then strain and use the liquid to rub into the skin after bathing. (See Queen of Hungary's Water for another recipe also reputed to be that of the Queen).

Elder flowers added to steam baths will clear and soften the skin. Freshly crushed leaves or freshly pressed juice of lady's mantle (*Alchemilla vulgaris*) is helpful against inflammation of the skin and acne, as well as freckles. Externally used lime flowers (*Tilia*) stimulate hair growth and are a fine cosmetic against freckles, wrinkles, and impurities of the skin. Aloe vera is now widely used in many skin preparations, and jojoba preparations are also becoming popular.

You can grow the very versatile luffa gourd for washcloths and bath sponges. This beauty treatment is centuries old in the Orient.

Use the oil of sesame seeds or the juice of lemons and cucumbers to soften and whiten the skin. Wheat germ oil and liquid lecithin (from soya beans) are believed helpful against lines and wrinkles.

Another oil treatment is rose oil. Steep rose petals in a bland oil, such as mineral oil. The petals may also be put into a crock and covered with water. Keep in a warm place and a little oil will rise to the surface. Collect it on a piece of dry cotton and squeeze into a bottle. This oil is attar of roses and is the rarest, most costly fragrance in the world.

COTTON, ORNAMENTAL. These two-foot plants have pink buds and creamy blossoms, followed by big white bolls of cotton.

COWSLIP (*Caltha palustris L.*) MARSH MARIGOLD. The plant is found in marshy places along creek beds and ditches, in swamps and wet meadows. The large plant *ssp. asarifolia* is abundantly found in southeastern Alaska and the coastal areas of the Gulf of Alaska westward to the Aleutians. A much smaller and less leafy plant *ssp. artica* is found throughout the Yukon and Tanana River basins. The bright yellow flowers may be borne singly or in clusters. The leaves and thick fleshy, smooth, slippery stems are best when young and tender before the flowers appear. The raw leaves contain the poison helleborin, which is destroyed in cooking. The roots are long and white. When boiled, the usual method of preparation, they look somewhat like sauerkraut.

CRAPE MYRTLE (*Lagerstroemia*). Masses of spectacular flowers make this shrub a southern favorite; it is dramatically beautiful when in bloom in the summer. And it comes in lovely colors of white, pink, watermelon red, two-tones, and royal purple.

Crape myrtle stands heat and drought well, and is not only easy to grow but easy to root. Cut off a branch, strip the lower half of leaves, and insert the cutting (or clone) in moist soil.

CREEPING JENNY (*Lysimachia nummularia*) CREEPING CHARLIE, GROUND IVY, GILL-OVER-THE-GROUND, CATSFOOD. This member of the mint family is a ground-hugging vine which returns year after year to produce pretty purple flowers and an aroma that protects nearby plants from insects. It spreads rapidly, sending down roots wherever it touches the ground, and likes a partially shaded location. According to some sources, sniffing the crushed foliage relieves headaches, and the roots contain a substance that stops bleeding.

CROP ROTATION. Crop rotation in the flower garden cuts down on certain insect pests. Spatial relationships are also important among flowering plants because crowding will reduce vigor.

CROSS-POLLINATION. Seeds may be produced by either self- or cross-pollination. In the former case only one plant is involved; in the latter, two. Pay attention to your nursery catalog to know whether you will need two plants for fertilization to occur. Wayside Gardens, for instance, advertises its "Breakthrough Hollies," "China Girl" and "China Boy," as a "fruiting pair." If you have a single holly, don't discard it; buy the missing member of the pair and your holly will bear berries.

Bayberry (*Myrica pennsylvanica*), beloved of early Americans for candle making, will not bear its berries unless staminate and pistillate forms keep company together.

The willowy leaves of the spiny sea buckthorn (*Hippophae rhamnoides*) are a lovely silvery gray, and pollinated female bushes bear an abundance of orange berries. No need to arrange individual marriages; plant one male buckthorn amid a small harem of ladies.

CROWN VETCH (*Coronilla*). Crown vetch is wonderful for preventing soil erosion along steep banks or for choking out unwanted weeds. It increases quickly by sending out roots both above and below ground. The neat, billowy foliage grows no higher than two feet and, in season, is a mass of lovely pink flowers.

CUCKOO FLOWER (*Lychnis floscuculi*) MEADOW PINK. Originally from Asia Minor and Siberia, this plant prefers moist meadows and has some value for feeding livestock. Its blossoms are bright red (the well-known garden varieties are pink, white, or blue, known as phlox).

The roots of all *Lychnis* species contain saponin, which produces a soapy foam if stirred in water. Before the discovery of soap, it was used together with the true *Saponaria* for washing. To this same group also belong: red campion, found on grainfields and in pastures, and white cockle or evening lychnis, so called because of its white blossoms which open in the evening and close at sunrise.

CULTIVATION. In preparation for planting, the soil of the flower bed should be in a good state of tilth with plenty of organic matter added for those flowers that require it, preferably in the form of compost. Cultivation is particularly important in the South and Southwest where the hot sun tends to burn the humus out of the soil. Replenish the bed with compost each season.

A small area may be spaded; for larger flower beds a rotary tiller is invaluable. The type with the tines to the rear is easy for elderly persons to handle, permits deep or shallow cultivation, and adds greatly to the pleasure of gardening.

CUPID'S DART (*Catanche*) LOVE PLANT. The flower spikes, springing from silver green foliage are an exquisite cornflower blue and stay in bloom from early June right up to late September. These are superb as cut flowers either fresh or dried to preserve the glorious color in long-lived arrangements. Plant in full sun in well-drained soil; they will seldom need dividing.

CUP PLANT (*S. perfoliatum*). This plant has yellow flowers, and its leaves join together to make a cup around the stem.

CUTTING GARDEN.   In an out-of-the-way corner plant a cutting garden so that there will be fresh flowers for the house without the necessity of robbing the display flower beds. No need to pay lots of attention to design or aesthetics—simply grow neat rows of those annuals that bloom abundantly in colors and forms you want for decoration.

If your color scheme calls for pinks and blues, raise larkspur, Canterbury bells, asters, bachelor buttons, felicia daisies, or stock. Add some dusty miller for its gray foliage; it's most compatible with pink tones.

For vivid reds, yellows, and oranges, grow marigolds, plumed celosia, red hot poker, geraniums, gloriosa daisies, and gazanias. The taller varieties are best. Coleus make a fine foliage filler with these flowers.

Poppies, both the Shirley and the Iceland poppies, are great additions to mixed bouquets. They are long lasting if you sear the stem ends when you cut them.

For airy fillers, grow some annual baby's breath or dill. And plant a few rows of strawflowers, statice, and bells of Ireland to cut and use fresh or dried in your home during the winter.

If you only have space for a tiny cutting bed, try tall zinnias and snapdragons. Their white, yellow, orange, red, and pink colors blend well and their forms contrast nicely.

CUTTING POINTERS.   In a pamphlet (Ext. Bul. 1192) R.T. Fox and J.W. Boodley of Cornell University give the following advice:

1. Cut the flower stems. A freshly cut stem absorbs water freely. Use a sharp knife or shears and cut either on a slant or straight across.
2. Follow special procedures for special cases. Some flower stems exude a milky fluid which plugs their water-conducting tubes. To prevent this, place about ½ inch of the stem in boiling water for thirty seconds, or char the end of the stem in a flame.
3. Remove excess foliage and those leaves that will be below water. Excess foliage increases water loss and submerged foliage decays and hastens fading.
4. Place the stems in water of 110° F. Warm water moves into the stem faster and more easily than cold water.
5. Use a commercial flower food in the water. These foods combine sugars, acidifiers, and a mild fungicide which lengthen the life of cut flowers.
6. Wrap a piece of paper or plastic around the flowers after you have put them in warm water. This will prevent rapid air movement over the flowers and reduce water loss. After the flowers become crisp (about two hours) you may arrange them and they will continue to take up water. Sometimes wilted flowers can be restored by repeating treatment.

7. Wash vases with soap and water after each use to remove bacteria. When bacteria multiply, they clog the water-conducting tubes of the flower stems.
8. Avoid excessive heat. Do not place flowers in direct sunlight, over a radiator, or in a draft.
9. Double the life of your flowers by placing them in a cold room or refrigerator at night, or when you are not at home.
10. Do not mix flowers with fruits or vegetables. Many fresh fruits and vegetables produce enough ethylene gas to shorten flower life.

CUTTINGS.   These are vegetative portions of plants used for reproduction. A cutting may consist of the whole or part of a stem (leafy or non-leafy), leaf, bulb, or root. A root cutting consists of the root only; other cuttings have no roots at the time they are made and inserted.

To assure success, make cuttings when the tissues are in the right condition. Then properly prepare and insert in the right rooting medium, and keep in a favorable environment until they are able to regenerate themselves as new plants.

Take a cutting of the plant part to be used of the correct size so the regeneration of new parts is encouraged; this may mean the removal of leaves, or parts of leaves.

Many different rooting media are used for cuttings. One of the most popular is sharp sand, or sand and peat moss, or vermiculite and sandy soil. Cuttings of some plants such as the African violet may be rooted in water.

For plants in all rooting media, prevent the tissues from drying before the new plant is established. This avoids loss by disease and encourages rapid re-establishment. The introduction of special root-inducing hormones and the use of bottom heat may speed this process. In the greenhouse this is usually done by using hot-water pipes, electric heating cables, or fermenting manure.

CUTWORMS.   Climbing cutworms crawl up the stems of plants at night to feed, so just because you don't see them doesn't mean they aren't there! Tansy is repellent to cutworms. Or collar stems of newly set plants, letting the cardboard collar extend both above and below soil line for about two inches. Occasionally a plant will be cut off even inside the collar; if this happens, dig down and find the cutworm before placing another plant.

A mulch of oak leaves or tan bark placed in strips in the beds and spread on garden paths will repel cutworms, slugs, and the grubs of June bugs.

D

DAFFODILS *(Narcissus pseudo-narcissus).* Daffodils announce the advent of spring. Their cheerful yellow combines well with the violet purple of the grape hyacinth, or you might finish the bed off with a ring of crocuses. In Europe and England the interest in daffodils has sometimes assumed the proportions of a craze. Rival enthusiasts grow and cross daffodils, getting extravagant prices ($500 to $2,000) for no more than five or six bulbs. In America we have never taken these flowers as seriously but love them nevertheless and plant them widely in our spring gardens.

DAHLIAS. If you want an ideal flower, try the vigorously growing dahlia for an abundance of beautiful flowers. And, best of all, dahlias are relatively free of diseases and pests. They come in many sizes, many colors, and single- or double-flowered. New cactus varieties are particularly attractive. Dahlias grow best in deep, fertile, well-drained soil in a sunny location; the plants are easily damaged by cold. Separate root clusters and plant about the time of the last killing frost. Space them three or four feet apart if you want large exhibition flowers. For large blooms allow only one stalk per root to develop. Remove all small weak sprouts. When the shoot is about six inches tall, pinch it back to the third set of leaves to promote branching.

Dahlias protect nearby flowers against nematodes.

Daffodils herald spring with their bright yellow blossoms. They'll also discourage moles. Daffodils and grape hyacinths combine well in the spring garden.

The black-eyed Susan is one of the nation's favorite wildflowers, growing throughout the nation, and even into Mexico and Canada. Easy to grow, it's an excellent cut flower.

DAISIES, WHITE or OX-EYE *(Chrysanthemum leucanthemum).* These are the well-known flowers with long stems, white and yellow centers, which often infest pastures, hayfields, and lawns. Often they are planted with grass seed but should be avoided in the lawn. They increase with increasing acidity of the soil, standing surface moisture, and loss of lime. Good neutral compost with lime in it, bone meal, surface harrowing or frequent raking of the lawn so the upper crust and root felt are broken, will take care of this problem.

Cultivated daisies are something else again, and there are many beautiful varieties for beds and borders.

DAMIANA *(Turnera aphrodisiaca).* This Mexican and African plant has been widely recommended by laymen and scientists for treating impotency and for its tonic effect on the nervous system. Long ago, the Aztecs used the leaves of damiana as an aphrodisiac. Steep one or two heaping tablespoons for five minutes in a pint of water. A commercial liqueur made from damiana, called Liqueur for Lovers, can be purchased in the United States.

DANDELION *(Taraxacum officinalis).* Dandelions, even if they grow in thick patches on your lawn, are not competing with grasses because of their long, very deep taproots. These transport minerals, especially calcium, upward from deeper soil layers, even from beneath hardpan, which they can penetrate, and deposit them nearer the surface. They are therefore returning to the soil minerals which have been lost through seeping downwards.

Dandelions are closely associated with clover and alfalfa, which also prefer good, deep soils.

Juliette de Bairacli Levy in her *Herbal Handbook for Everyone* states: "This is one of the most esteemed plants of the herbalist. . . . It is blood-cleansing, blood-tonic, lymph-cleansing. Also has external uses for treatment of warts and hard pimples. A diet of greens improves the enamel of the teeth."

Dandelion helps other flowers grow. It stimulates fruits to ripen faster. The roots, dried and ground, are sometimes used as a coffee substitute. In early spring the unopened buds are delicious cooked with leeks, lightly seasoned with butter, salt, and pepper.

A pale yellow dye can be made from dandelion blossoms; a deeper yellow brown color is obtained when the dye is made from the plant's roots. An improved, thick-leaved dandelion variety with dark green leaves is available from the Burpee Seed Company. The hearts may even be eaten raw if the leaves are tied together and the hearts blanched. They are very rich in vitamin A.

DAPHNE ODORA *(D. indica).* This small evergreen shrublet has white or purplish flowers in January. It is said the daphne "can boast of being the most powerfully fragrant plant in the world." It grows as far north as Washington, D.C., and persists over winter if given a warm wall to sun its back against.

D. laureola (spurge laurel) grows luxuriantly in shrubberies where it is hardy and often produces its small green flowers as early as January. The plants have a delicious scent like primroses which can be detected at a distance of thirty yards.

D. mezereum, however, produces poisonous berries, and eating even a few can be fatal to a child. It is often grown as an ornamental. Its fragrant, lilac purple flowers in stalkless clusters of three bloom before the leaves come out.

DATURA *(Datura stramonium)* THORNAPPLE, JIMSONWEED, NIGHTSHADE FAMILY. Both seeds and leaves yield narcotic drugs used in medicines. The strong odor of the plant causes drowsiness and if used sparingly, smoke from dried datura leaves is calming to honeybees when opening a hive. Sucking nectar from the flowers has poisoned children, and some have died from swallowing the seeds. Even eating the boiled plants produces irrational behavior. Early California Indians knew this and gave their children potions made from the plant to obtain visions especially upon reaching puberty, but expert shamans regulated the dosage.

Datura, though somewhat coarse, is nevertheless a beautiful plant and because of the large white blossoms is also called angel's trumpet. Datura helps the growth of pumpkins when planted in their vicinity.

DAYLILY *(Hemerocallis).* The daylily has long been a mainstay of perennial plantings. The usual method of propagation is by clump division, easily done in autumn, but for best results a clump should not be divided more often than every two or three years.

Another reproduction method is by proliferation. The sturdy little rootless plants that develop along the flower scapes (stalks) in the axil between the rudimentary stem leaves and the stalk eventually die if they do not touch ground; nature probably intended them to spread the plant when the flower stalk finally breaks and falls to the ground. Give nature an assist. When the flowering period is past and the stalk starts to dry, cut off the proliferation by severing the stalk about two inches below. Insert the proliferation in a pot full of good growing mix with the base just below the soil surface. Keep soil moist. Roots will soon develop from the base. After good root development, set the plant out in the garden.

DEATH CAMAS *(Zigadenus venenosus)* LILY FAMILY. Sometimes cultivated in gardens, this bulbous perennial plant is found below 8,200 feet in meadows, pastures, open slopes, and along roadsides from Canada to Florida, Texas, New Mexico, Arizona, and California. All parts of the plant are toxic; poisoning has followed even the eating of the flowers. The onionlike bulb has a dark-colored outer coat but lacks the onion odor.

Another variety of camas, the Indian quamash, "Queen of the Bulbs," played a vital role in the history of the Indians. With breadroot and cous, camas constituted their basic starch food.

Camas of this type are found as high as the subalpine, growing along streams or in moist meadows. They cover vast areas so closely that in springtime the effect is that of lakelets of brilliant blue. Their flowers, grouped on spiny racemes, are spectacular. The ovate bulbs also look like onions.

The Indians learned how to differentiate between the two—the quamash with its big blue flowers, the "bringer of life," and the death camas, the "bringer of death," with its small yellow blossoms.

DECOR-EDIBLE VEGETABLES. A garden of decorative vegetables may be planted alone, or the vegetables interplanted in the flower garden. Highly attractive vegetables, both in form and color, add a note of unusual interest. Consider the climbing, or tree tomatoes, for a trellis— or a planting of cucumbers, which will grow straight up if supported. For a border, plant Tiny Tim or Small Fry (red), Yellow Tiny Tim, or red or yellow cherry tomatoes. Eggplant comes in white or apple green as well as the shiny purple black most often seen; it also is attractive when in bloom. Broad beans have showy flowers, as do scarlet runner, edible-pod sugar peas, and asparagus peas. Vivid green parsley makes a nice low edging.

Red okra is red all over, plant and pods, though they turn green when cooked. The pods are beautiful for dried arrangements; if allowed to mature on plant, they become a rich brown with sculptured looking shapes.

Bush summer squash has beautiful flowers and foliage with fruits of interesting shapes and colors. Many pepper varieties are exceedingly pretty and change color as they ripen. There are also the small, ornamental hot peppers.

Crinkly kale is a pleasing border plant. So are the red cabbages and the savoy types. So-called "flowering" kale and cabbage always attract notice when they hit their color peak in fall. They are called flowering, not because of their flowers but because of their colorful crinkly leaves which come in creamy white or reddish tones against green. Ruby lettuce is another beauty not to be overlooked.

DELPHINIUM *(Delphinium spp.)* Larkspur, Crowfoot family. Widely cultivated for their beauty, many of these plants have escaped to roadsides and fields. The prevailing color is blue, but cultivated forms

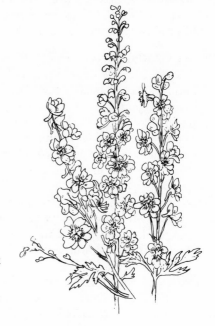

One gardening book calls the delphinium "a confused hybrid of uncertain origin." No matter, it's a hardy flower that's easy to grow, and offers blossoms that are pink, blue, purple, or white.

come in many colors—some even have beautiful double blooms. All species of the plant contain alkaloids of varying quantities. Ingesting young leaves before the flowers appear causes poisoning, but the plants' toxicity decreases as they age. Leaves and seeds may cause dermatitis on contact.

Common field larkspur (*D. consolida*) yields the alkaloids delcosine and delsoline, found effective against aphids and thrips. Powdered roots are toxic to bean leaf roller, cross-striped cabbageworms, cabbage loopers and melonworms.

DENTARIA.   The name is from the Latin *dens*, a tooth, and refers to the toothlike scales of the roots. The plants are uncommon but useful little perennials for shade. Try them in porch boxes located on the north.

DERRIS.   Long ago the Chinese discovered the insecticidal value of the derris or tuba root. It can be used on both plants and animals for insect control. It is notably effective against leaf-eating caterpillars, mosquito larvae, and aphids. The addition of oil of teaseed as a synergist increases its toxicity to squash bugs.

Derris is widely used on farms as animal dips to control lice and ticks. Applied to young bean plants, it makes the leaf growth less palatable to Mexican bean beetles.

When the crushed roots, stems, and leaves are thrown into lagoons and streams, fish float to the surface, insensible. Despite the extreme sensitivity of the fish, however, derris in normal concentration is believed harmless to domestic animals and man.

DESERT GARDENING.   In the desert the lawn area is kept small, and generous use is made of rocks, gravel, and patio paving. Raised beds, pools, and water plants are often found. Trees that can "take it," such as cottonwoods, poplars, Siberian elms, black locusts, chinaberry trees, and evergreens such as athel tamarisk, Arizona cypress, and Aleppo pine, are carefully chosen for shade and wind control. In low deserts this list should include more evergreens, such as beefwood, black acacia, and eucalyptus. Survival and fast growth are important criteria for desert plants.

Good companion plants are night-flowering cactus, Mexican fan palms, yucca, and low-growing lavender lantana.

For easy upkeep annuals and perennials such as creeping rosemary, feathery wormwood, senna, African daisy, salvia, and fairy primrose are grown in flower beds.

For seasonal color plant pyracanthas, oleanders, yuccas, and showy crape myrtle. Remember: if plants can't take heat, drying winds, and alkaline soil and water, their beauty counts for little.

DICTAMNUS. Dittany, Burning Bush.   The ancient Greek name, *dictamnos*, refers to Mount Dicte, Greece, where, according to legend, Zeus was born, and where this plant once grew. One species, a curious perennial, gives off a volatile oil from the upper parts of its stem which may be ignited in hot weather and will burn without harming the plant.

DIERAMA. Wand Flower.   From the Greek *dierama*, a funnel, the name describes the shape of the individual flowers hanging from long, slender stems. These perennials from South Africa cannot withstand wet, cold winters; use as houseplants in northern areas.

DOCK, ARCTIC *(Rumex)* Sourdock, Wild Spinach.   The flowers are green or tinged with purple, numerous, and mostly crowded in panicled racemes. The plant grows in wet, marshy places along river banks in Canada and Alaska. Its young tender leaves, an excellent source of vitamins A and C, make an excellent salad green and cooked vegetable.

DOG FENNEL *(Anthemis cotula).*   This plant is sometimes mistaken for camomile, but once you know the true camomile fragrance you will never confuse the two; dog fennel has a rank, weedy odor. To tell them apart, cut one of the well-developed, little daisylike blossoms vertically through the middle. True camomile has a hollow center; dog fennel is solid.

DOG-GONE!   Man's best friend can be a nuisance sometimes. He can leave yellow patches on the lawn, ruin shrubs, and dig in your best flower bed. Sprinkle mothballs or naphtha flakes around the beds to discourage him. Or pound small sticks a foot or so apart in the region you're trying to protect.

DOGWOOD *(Cornus nuttallii).*   The western dogwood, a very handsome tree, has beautiful blossoms and an agreeable honeylike fragrance. *C. amomum*, the red-stemmed dogwood, has fragrant inner bark which the Indians use for smoking.

DOLLS.   In Colonial times corncob dolls were enchanting little toys that children loved. Shell off the corn, dry the cob with the shucks left on, then pull the shucks up and back for the "hair." Slit the shucks and make them into braids. Or they may be curled by holding a strip of slit shuck between thumb and a knife and pulling the shuck through, just as ribbon is curled. Try decorating a small Christmas tree with tiny dolls made from midget corn.

When digging potatoes, watch for those with unusual shapes; they often lend themselves to roly-poly comic characters or animals. Carrots with a little trimming can be made into dolls as well. Dolls with apple heads have been made for centuries; some of them are truly works of art.

DOWSING. To find out if you are a water witch, take a small tree branch with a fork in it—a branch shaped something like a wishbone. Hold the two ends of the forked branch with the palms of your hands facing up. Walk slowly with the branch pointing straight up. When you approach a source of moving water, even through an underground water pipe, you will feel a definite pull on the branch as it begins to move toward the earth. When you are directly over water (*if* you have the power of a water witch), the branch will insist upon pointing straight downward even though you are trying your best to hold it up.

Water witches throughout history have been called upon to use their divining rods; those who are known to be able to find water are still called upon. Some can even get results by holding the rod over a map of the area.

Most water witches say that a forked stick from almost any tree will work if you have the power, but the preferred species are peach, apple, willow, and maple.

DRAGON'S HEAD (*Dracocephalum*). From the Greek, *drakon*, a dragon, and, *kephale*, the head, the name refers to the gaping flower mouth. Both annuals and perennials are useful for the front border.

DRIFTWOOD. Often free for the gathering, sculptured by wind and wave, driftwood makes a beautiful "pot" for cactus, sedums, and other small plants. Or fill it with lacy ferns with their own peaty earthballs in plastic bags, punctured at the base for drainage.

DROUGHT-RESISTANT FLOWERS. Cornflower, calliopsis, sunflower, morning glory, ice plant, four o'clock, rose moss, and zinnia will grow in regions with deficient rainfall.

DRYOPTERIS AUSTRIACA. Mainly found in moist woodlands in southeast Alaska and southward, this lovely, spreading woodfern is delightfully edible. When the fern fronds or blades first appear, they are curled and chaffy. These are called croziers or fiddlenecks. The spores are borne in "sori" on the back of the frond. They are small, round, brown, and set in from the margin of the frond. The old leaf stalks on the underground stem, resembling a bunch of minute bananas, have been used for centuries by the Indians. They are roasted, then the outer shiny brown covering is removed and the inner portion eaten. They are a fine source of energy. The young curled fronds, when still only about five to six inches high, are collected in the spring. They are boiled or steamed and served like asparagus, either with butter, margarine, or cream sauce. Many southeast Alaskans can them for winter use.

DUCKWEED (*Wolffia punctata*). Duckweed is the smallest known flowering plant, and grows about $1/50$ inch long and $1/63$ inch wide. Duckweeds seen on the surface of ponds are rootless.

DULSA *(Rhodymenia palmata)*.   This is a seaweed that consists of thin, elastic lobed, purple red fronds varying from a few inches to a foot in length. It is found attached to rocks or coarser seaweed near the low tide mark along the Pacific Ocean shore of southeastern Alaska. The southeast Alaska Indians gather it in quantity in spring, air-dry it, and store it for winter use. It is often added fresh to soups and fish head stews. It contains iodine.

DUMBCANE *(Dieffenbachia)* ARUM FAMILY.   These evergreen foliage plants are widely grown in greenhouses, homes, restaurants, and lobbies as potted ornamentals. They thrive planted outdoors in the southern part of the United States. The flowers are tiny; the fruit is fleshy.

All species contain calcium oxalate needlelike crystals in the stems and leaves. The two most commonly cultivated species are *D. picta* and *D. Seguine*. It is called dumbcane because chewing on it causes temporary speechlessness.

DUTCHMAN'S BREECHES *(Dicentra cucullaria)*.   This plant is a native of the eastern woods and has a surprisingly sweet scent. But beware: Eating the leaves and roots produces poisoning and such nervous symptoms as trembling, loss of balance, staggering, weakness, difficulty in breathing, and convulsions.

DUTCHMAN'S PIPE *(Aristolochia durior)*.   This foliage vine with its handsome, heart-shaped leaves creates dense cooling shade. It grows rapidly, reaching up to thirty feet in height.

E

EARTHWORM.   The castings of the earthworm are rich in nitrates, phosphates, potash, and calcium—all elements necessary to plants. They also contain trace elements of sulphur, boron, zinc, copper, manganese, chlorine, iron, molybdenum, aluminum, and selenium—all needed to keep a plant healthy. Gardeners everywhere are beginning to realize the value of earthworm castings. They report that their use results in vegetables of better taste and larger size with yields often doubling. Tomatoes especially benefit from castings.

Here are some suggestions:

Indoor Plants.   For established pots, add one cup of castings to a six- or eight-inch pot and water thoroughly. Increase proportionately for larger pots. Castings will be absorbed into the soil during normal waterings.

Sick or Dying Plants. Remove one to two inches of the old planting mix (being careful not to disturb the plant any more than necessary). Replace with castings.

Root Stimulator. Use castings to make up the planting root ball for newly planted fruit trees, rosebushes, and berry vines. This application of castings (where other fertilizers are too strong) provides a head start for the plants and will produce remarkable results even for years to come. Seeds also germinate much faster in castings.

Potting or Re-potting. Mix ⅓ castings with ⅔ peat moss, or any good soil. Add enough of this mixture to set plant at original depth. Pot plant, fill around it with the castings, and press gently to firm the soil. Water thoroughly.

Castings will not burn even the most delicate plant. There is no danger of adding too much to your soil. Worm castings can be found packaged, ready to use, in your local supermarket.

EAU DE COLOGNE.   This spirituous preparation contains oils of rosemary, citron, orange, bergamot, neroli, and geranium. It has been produced since 1700, and is very useful for freshening a sickroom.

EDELWEISS *(Leontopodium alpinum)*.   This hardy perennial likes sun and a dry location. Do not cover the seed because light promotes germination which takes place in fifteen to twenty days. The gathering of edelweiss has long been symbolic of daring achievement, since it is native to the high rocky ledges of the Swiss Alps. The wooly, star-shaped flowers are snow-white at high altitudes; at lower elevations, they are a soft gray green. The plant grows twelve inches high. (Applewood Seeds)

EGLANTINE *(Rosa rubiginosa)* SWEETBRIER, WILD ROSE.   Eglantine is lovely but can become too much of a good thing. It immigrates from hedgerows to pastures where it shoots up quickly. This indicates that the pasture has not been grazed sufficiently and should be mown and harrowed. Due to their prickly canes they can be troublesome to cattle and sheep. If they become established, they will protect the growing and seeding of other weeds. Cut while still young when canes are soft.

ELDER *(Sambucus)*.   Elder is noted for repelling certain insects. An old method of trapping cutworms consisted of placing compact handfuls of elder sprouts in every fifth row or hill of cultivation and tamping them down. Cutworms gathered in this trap material where they could be regularly collected.

Elder is a powerful patriarch of the plant world; wherever it grows it discourages other herbs. Animals usually dislike the rank elder taste.

Elder leaves are effective against moles if placed in their runs; they find the odor offensive. Branches of elder have been used against maggots. Bruise them first to increase the odor, then rake the leaves across the seedbed after sowing.

ELEPHANTHEAD. This large clumping perennial grows to three feet with delicate fernlike leaves. The rose pink flowers resemble an elephant's head with its large floppy ears and long trunk. Flowers in tall spikes emerge from the ferny foliage in August. The plant is nice for naturalizing as it grows well in moist meadows, bogs, stream and lake shores from Greenland to Alaska and south to the mountains of New Mexico. Seed should be cold stratified for thirty days before sowing in spring. (Plants of Southwest)

EQUISETUM. According to Pfeiffer in *The Pfeiffer Garden Book*, the biodynamic remedy for fungus diseases and rusts, other than treatment of the soil, lies in a spray made from the common horsetail, *Equisetum arvense*. The spray consists of the following: "Take one and a half ounces of the dried herb and cover with cold water. Bring to a boil, and let boil twenty minutes. Cool gradually, then strain. Use one part of this concentrate to nineteen parts of water as a spray." This tea is also useful for treating sores on domestic animals.

This interesting plant is the last reminder of the huge trees of the carbon forests; it propagates by spores and creeping rootstocks. All equisetum plants, especially the field horsetail, have a silica skeleton. Before cleaning agents were manufactured, boiled horsetail was used to clean and shine surfaces of silver and pewter.

ESPALIER. For training small plants as miniature espaliers, try the annual balsam. Pinch off side shoots so it will grow tall and slender; you will be rewarded with a mass of blossoms. Espalier form is still used for training fruit trees on frameworks or against a wall, although it is not as popular as it once was.

ETAK *(Eriophorum augustifolium)* Eетаht, Tall Cottongrass. The flower heads of this perennial develop into two to twelve nodding heads of white, silky bristles called cotton by Alaskans. Etak is found on tundra bogs and wet roadsides. In autumn tundra mice cache the underground stem for winter use. The Eskimos call these underground stems "mouse nuts." Sometimes they eat the nuts with seal oil.

EUCALYPTUS *(Eucalyptus)* Gum Tree. This tree has a remarkable capacity for storing solar energy. Experiments in South Africa have shown that a forest of such trees produces yearly approximately twenty tons of fuel per acre. The dry timber is heavier than coal and gives out as much heat when burned. These trees thrive best in hot, moist regions, but some varieties are extremely drought-resistant.

Young blue gum *(E. globulus)* is handsome for a houseplant or for planting in the garden for summer foliage effects. Foliage of most eucalypti is fragrant. The lance-shaped leaves are long, narrow, and leathery.

The feathery flowers look like bells and are filled with nectar. In California, eucalypti are planted around orange and lemon groves to break the wind. The trees also furnish a resin, called Botany Bay kino, which protects wood against shipworms and other borers. The bark of some species furnishes tannin, which is used medicinally. Eucalyptus leaves contain a valuable oil that smells somewhat like camphor and is used as an antiseptic, deodorant, and stimulant.

Gather bark, stems, leaves, and seeds of long leaf eucalyptus and make a decoction by boiling. Use to spray plants affected with aphids.

EUGENOL.   A chief constituent of oil of cloves, eugenol is an effective attractant for baiting insect traps. Star anise and citronella grass are also useful for this purpose.

EUPHORBIA (*Spurge*).   This plant has been rated by one of the world's outstanding nursery experts as among the ten best perennials for its long-lived reliability, ease of cultivation, neat impressive form, and outstanding color. It looks a bit like cactus but is totally unrelated. Shapes run from clean geometrics through organ pipes, fat balls, and cylinders. Most strange of all are those with convoluted crests and snakelike, Medusa head forms.

All euphorbias have small flowers without petals, enclosed in a cup-shaped, leaflike structure with five lobes and a honey secreting gland. The single female flower is normally surrounded by numerous male flowers.

EVENING PRIMROSE (*Oenothera*).   Oil of evening primrose is said to be the world's richest source of natural, unsaturated, fatty acids. It is helpful in cases of obesity, mental illness, heart disease, arthritis, and for relief from postdrinking depression.

The cultivated *O. tetragona* (fireworks) is a cheerful plant with dark green dwarf foliage, tinted purple in the spring. The red buds open to profuse silky bright yellow flowers borne during June and July.

Oenothera missouriensis, evening or Missouri primrose, is classed as a wildflower. It has immense four to five-inch, cup-shaped, yellow flowers. A low-growing species, it is ideal for border or rock garden. The winged seed pods are tan, often streaked with crimson, and are excellent for dried arrangements. The plant likes fertile, well drained soil and a sunny location.

EVERGREENS.   Needles are useful for soil building and make good humus for azaleas. Evergreen plantings make good windbreaks.

EVERLASTING, FRAGRANT (*Gnaphalium polycephalum*).   This is the fragrant immortelle of the autumn fields, spicy and sweet and often growing in old fields and woods. *G. ramosissium* is a fragrant pink flowered everlasting.

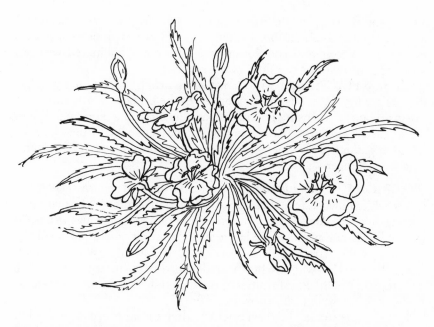

The showy evening primrose has day-blooming relatives called sundrops. It is easy to grow, easy to divide. The oil of evening primrose from it is used for a multitude of health problems such as heart disease and arthritis.

EVERLASTING PEA.   Use this plant as a control against field mice.

EXOTIC FRUITS AND PLANTS (*Actinidia chinensis*) Kiwi.   The so-called Chinese gooseberry is a native of China and is commercially grown in New Zealand. The plant is a rampant grower, growing possibly five inches in one day and up to eight feet the first year. The fruit tastes like a blend of strawberry, pineapple, and guava. Lakeland Nurseries offers the Hayward variety said to have especially large fruits with fuzzy brownish skin, eye-appealing green flesh, and edible seeds. It also produces showy, camellialike flowers in spring.

Train grapelike vines on arbors, trellises, or fences for November harvesting. The fruits keep for eight months or more. In northern states they may be grown in tubs and should be brought indoors if temperatures drop below 20°F. Grow in pairs of one male and one female vine; additional female vines may be grown with one male.

Cherimoya (*Annona cherimoya*). This small, unusual, tropical American tree grows wild in Peru and is now cultivated in California and Florida. The tree bears fragrant yellow flowers followed by egg-shaped or heart-shaped fruits weighing a pound or more. Its white smooth pulp tastes like a mixture of pineapple, peach, and banana. According to

Thompson & Morgan, the tree grows quickly and has very ornamental foliage; the fruit is about four inches across. However, it normally takes two years to fruit and the flowers must be hand pollinated with an artist's paintbrush for the flowers to set.

Chinese Fly Catching Vine *(Aristilochia delibis).* The long peculiar flowers are insectivorous; their odor serves the purpose of attracting the insects required to ensure pollination and fertilization. The small leaves are attractive, and the plant has hanging-basket-shaped fruits. Use for hanging baskets. This vine is an important medical plant in China and Japan. (Mellingers)

EYEBRIGHT *(Euphrasia officinalis).* Eyebright is characterized by lovely white flowers striped violet which open in summer. The plant was once thought to be effective in treating opthalmia. Culpepper writes: "If the herb was but as much used as it is neglected, it would half spoil the spectacle makers' trade. The juice or distilled water of eyebright, taken inwardly in white wine or broth or dropped into the eyes for diverse days together helps all infirmities of the eyes that cause dimness of sight. It also helps a weak brain or memory."

Some people smoke dried eyebright as they would tobacco. A kind of wine can be made by cooking it with unfermented wine at harvest time; however, eyebright should not be taken internally because there have been cases of stomach disorders. The juice is acrid and has an unpleasant taste.

EYES. Various Indian tribes developed herb and flower remedies for sore or strained eyes. Indians often named their plants according to the way they were used. "Put into Eyes" is the Indian name for the prairie zinnia *(Crassina grandiflora).* The Zuni crushed the flowers in cold water and used the strained liquid as an eyewash. "Weak Eye Tea" is an Indian name for the wahoo *(Euonymus atropurpureus).* Several tribes made infusions for sore eyes from either St. John's-wort or St. Andrew's-cross. Other mild infusions were made of any one of the following: alder bark, bearberry leaves, yarrow herb, chickweed herb, blackberry root, black oak bark, choke cherry bark or bark, of button bush. Ginseng root was also used for sore eyes, the pounded root being soaked in cold water and then strained.

Buds of sassafras were gathered by southern Indians and placed in cold water, the mixture then was allowed to stand several hours in the sun. The glutinous substance formed was used in treating sore eyes.

Herbalist Jethro Kloss recommends a poultice of slippery elm applied (cold) to the eyes to relieve inflammation. Other herbs he recommends for sore eyes are rosemary, borage, camomile, chickweed, elder, fennel, goldenseal, hyssop, rock rose, sarsaparilla, sassafras, witch hazel, wintergreen, yellow dock, plantain, tansy, white willow, and angelica.

F

F_1 HYBRIDS, F_2. Burpee's mule marigolds are F_1 hybrids. If mules could reproduce, their offspring would be F_2 hybrids. The "F" stands for filial or offspring, the digit for the generation. Just as mules have greater vigor than either parent, so have hybrid plants. In general, they grow faster, bigger, and more uniform, and bear more flowers and fruit. That's especially true of the F_1 generation. Offspring of F_1 plants—the F_2 generation—have more vigor and uniformity than regular plants, but less than the F_1. F_1 hybrid seed is expensive because seedsmen can't let bees do the pollinating, but must brush the proper pollen on by hand. F_2 seed doesn't need hand pollination, so is less costly.

FALSE SAFFRON. SAFFLOWER.   False saffron is a thistlelike plant with large, attractive flower heads varying from white to brilliant red. It has been grown in dry areas of Asia, Africa, and Europe for hundreds of years. The chief value is in the oil and meal made from the seeds. The oil is used in the diets of persons suffering from heart disease and hypertension, and is a valuable source of linoleic acid. The meal is fed to livestock.

FATIGUE.   The Aztec Indians breathed flower fragrances to ease fatigue. They also recognized melancholia and loss of memory as diseases. The fragrance of flowers was used freely for its psychic effect; a lotion compounded of flower concentrate and other ingredients calculated to retain the delicate aroma was used in body massage. A similar preparation of flowers was also taken internally.

FENCE.   Robert Frost said, "Good fences make good neighbors." They also keep out unwanted animals. A living fence such as multiflora roses may also act as a deterrent and give lovely bloom as well. And think of the vitamin C in all those rose hips.

FENNEL, SWEET (*Foeniculum officinalis*).   In the garden fennel is valued for its masses of fringed foliage. In times past the fragrant seeds were made into a tea for babies' colic. Mixed with peppermint leaves they also make a delicious tea, calming to the nerves. The Italians consider fennel valuable as a key to longevity and virility; the Egyptians, Greeks, and Romans all vouched for fennel, using it in salads, fish sauces, and fennel soups. Fennel oil, which contains estrogens, has a reputation as an anti-wrinkle agent.

The Florence fennel, prized for its enlarged leaf bases and used like celery, is a close relative and is grown the same way. These fennels are poor companions for bush beans, caraway, kohlrabi, and tomatoes. The black swallowtail butterfly lays its eggs on fennel, dill, and parsley. The beautiful black and gold caterpillars are ravenous.

FEVERFEW (*Chrysanthemum parthenium*).   Feverfew, an old favorite for edging, has little white buttons ¾ inch across, and yellow foliage with a characteristic strong, bitter odor.

This little daisylike flower is sometimes incorrectly called pyrethrum. However, like the real thing, it works as a bug chaser and can be planted as a border around roses or scattered throughout the garden. Feverfew grows in tufts, becoming bushlike and occasionally attaining a spread of as much as three feet. The single-flowered form was the feverfew cultivated in old physic gardens. The modern forms, largely used for cutting, are double-flowered. Feverfew is an accommodating flower, will grow in any soil, gives generous bloom all summer, and self-sows readily.

FEVERWEED (*Eupatorium purpureum*).   The Joe-pye weed grows mostly in moist damp thickets, ditches, and streams, invading only badly drained meadows. The plant is named for an Indian herb doctor, but is also called feverweed. The juice is said to heal open sores and bruises; hunters have observed that wounded deer will search for it and eat it. (Early peoples learned much about the medicinal qualities of plants by watching animals when they were sick or injured.) Feverweed is closely related to thoroughwort (*E. perfoliatum*).

Feverfew is great for a border planting, with its bushy growth and buttonlike blossoms. Old-timers believe the flower repels many insects.

FIG *(Ficus).*   Figs have their flowers inside the fruit. The fig wasp, blastophaga, lays its eggs at the base of the flowers, thus insuring pollination. Figs have long been considered love food by primitive peoples.

FINGERLEAF WEED.   This weed lives on acid soil and indicates increasing acidity. Weeds are specialists and close observation tells us a great deal about the soil they grow on.

FIRE-RETARDANT PLANTS.   While no plant will completely keep a fire from advancing, some plants resist burning better than others and thus may slow a fire's progress. However, if winds carry sparks, even protective fire-retardant plants can be breached. Useful trees and shrubs are: Callistemon, *Ceratonia slilqua, Heteromeles arbotifolia,* Myoporum, *Nerium oleander* (dwarf varieties), *Prunus lyonii, Rhamnus alaternus,* Rhus (evergreen kinds), *Rosmarinus officinalis prostratus, Schinus molle, Schinus terebin thifolius, Teucrium chamaedrys.* Perennials and vines: Achillea, Agave, Aloe, Artemisia (low-growing varieties), Atriplex, Campsis, *Convolvulus cneorum,* Gazania, ice plants, *Limonium perezii, Portulacaria afra, Santolina vivens, Satureja montana, Senecio cineraria, Solanum jasminoides,* Yucca (trunkless varieties).

FIREWEED *(Epilobium augustifolium L.).* GREAT WILLOW-HERB.   The large, showy flowers are borne in terminal spikelike clusters. They have four purplish or rose-colored petals which may be occasionally pink or rarely white. The plants are common along roadsides and on open hillsides from southeast Alaska north to the Arctic, and west to the Alaska Peninsula and the Aleutian Islands. They also are found in the mountains along streams and in clearings throughout North America and Eurasia. Fireweed was one of the first plants found in the Mount St. Helens area after the recent volcanic eruption.

The Indians collect the young shoots in spring and mix them with other greens; they peel and eat the young stems raw. Like other tender young greens, these are a good source of vitamin C and pro-vitamin A. Fireweed honey is one of the finest.

FIRST GARDEN BOOK.   Published in England in 1563, the first garden book, entitled *A Most Briefe and Pleasaunt Treatyse Teachynge Howe to Dress, Sowe and Set a Garden,* was written by Thomas Hyll, Londoner. "If you want your Parsley to be crinkled or curled," he writes, "bruise the seed, or when it comes up roll small weights on it, or else jump up and tread it down with your feet." Botany and medicine, which in Hyll's time were identical, were just beginning to free themselves from the influence of superstition and witchcraft—influences that are frequently evident in Hyll's book.

From it we learn much about gardens at the time of Queen Elizabeth

I. They were laid out formally with arbors and trellises. Mazes and knot gardens were popular, and often wells were included "for water is a great nourisher of herbs." Beds were raised for drainage, and walks were sanded "lest by rayne or showers the earth should cleave and clogge on thy fete."

FISHING WITH HERBS. Oil of anise rubbed on bait will attract fish, so will the juice of smallage or lovage, and the steeped root of sweet cicely.

FLAGSTONE FLATTERY. A flagstone walk gives interest and drama to the flower garden. To lay the stones in concrete, first wax the upper surface of each flagstone with liquid or paste wax. Cement will not adhere to a waxed surface, so it will be easy to clean off any spills or smears when the job is finished.

FLAX *(Linum).* Narbonne flax is an elegant, free flowering perennial with feathery, blue green, evergreen foliage, just right for a sunny spot or a rock wall. The flowers, 1¾ inches across, are sky blue with dainty white eyes. The 1½-foot plants with slender, erect stems bloom generously.

Dwarf golden flax *(L. flavum compactum)* has a myriad of bright yellow translucent flowers one inch across and makes a fine border plant. Give these full sun and a moist but well-drained location.

FLEABANE *(Erigeron).* This weed invades relatively good land, and is one of the few "presents" the American weed continent has given Europe. It was inadvertently introduced about 1655 in a stuffed bird. Its acrid oil is used against mosquitoes, hence the name "fleabane." Sensitive people may be allergic to this weed; however, it is still collected for medicinal purposes. Canadian fleabane *(E. canadensis)* can be ground up to make melonworm repellent.

FLEA BEETLE. This pest is repelled by wormwood, mint, catnip, or you might try interplanting susceptible crops near shade-giving ones.

FLORAL PRINTS. Gather thin, light, colorful flowers (pansies, petunias, columbine, buttercups) just before they come into full bloom. Brush off pollen and press between sheets of absorbent paper (such as paper toweling). Place a heavy object on top. Replace paper after eight hours and again after the next eight hours. Then leave for two to three weeks. Mount your floral prints on construction paper in any design you wish. No glue is necessary as the pressure of the glass will hold them in place after they are framed. Ferns also press well.

Pressed like this the flowers can be used to decorate Easter eggs. Touch them lightly on one side with Elmer's glue and gently press them on the egg.

Another method is to press them between sheets of plastic wrap and touch them lightly with a warm iron to seal. These make attractive greeting cards between two sheets of folded notepaper.

To make an ink print of dried ferns or flowers, use a blockprint roller. Ink the roller and roll back and forth over the materials to be used. Then place in desired arrangement on construction paper; roll clean roller over design. Different colors on the same arrangement give an interesting pattern.

FLOWERING TREES. Not to be overlooked are the marvelous flowering crabs. Some bear fruit but they are grown mostly for their beauty. Other flowering trees include: almond, cherry, peach, plum, and quince. Unequalled for its lovely display in early spring, is the cotinus, the so-called smoke tree. *C. coggygria* is Royal Purple indeed with its coppery purple black foliage and plumed inflorescence of the same color.

FLOWER POTS. Let your dishwasher sterilize your flower pots. First scrub your pots clean to remove soil and accumulated salts. Put clay pots on the bottom rack, plastic ones on top with pan lids on top to weight. Nest several sizes to maximize space. Pots will come out clean, sterilized, and ready to use.

FLOWERS, PROTECTION OF. Garlic planted with flowers protects against insects and often increases fragrance. Onions, chives, and leeks are also protective, as are the tall-growing decorative alliums.

FLOWERS, TOP TEN. The W. Atlee Burpee Company announced that the ten top flowers for which their customers buy seed are marigolds, zinnias, impatiens, petunias, geraniums, celosia, asters, snapdragons, coleus, and fibrous-rooted begonias.

The runner-up was starflower, or *Scabiosa stellata*. It has brown balls of spiky florets and is grown for dry-flower arrangements, crafts, or even Christmas tree ornaments.

Geraniums in fifth place is a surprise; just a few years ago they weren't even among the top ten because seeds weren't available. Today new varieties (such as the Sprinter hybrids) are practical to grow at home from seed, and they bloom much faster than the older types.

FLOWERS ON STAMPS. One of the six most popular design topics for postage stamps the world over is flowers. Switzerland was the first country to bring the beauty of a flower to a postage stamp. For many years Switzerland issued a yearly colorful series showing its native flowers: edelweiss, alpine rose, slipper orchids, and many others. Inspired by the Swiss success, other nations including the United States followed suit; now almost every country that issues stamps has honored flowers on some of them.

FLOWERS SENT TO HOSPITAL PATIENTS. Sending cut flowers to hospital patients recovering from surgery is a thoughtful but dangerous gesture. According to a British medical journal, it may lead to possible infection. A concentration of dangerous bacteria may grow in vases within one hour after flowers are put in water. And after three days, some of the bacteria are resistant to commonly used antibiotics. Flowers, the article continues, should be avoided in hospital units dealing with intensive care, burns, neuro-surgery, and newborn babies.

FOAM, PACKING. Save that packing foam next time you receive a package. Because it is lightweight, non-compacting, and water-repelling, packing foam is good material for hanging baskets and planters.

FOLIAR FEEDING. Spraying or applying fertilizer to plant foliage is just as effective with houseplants as with outdoor plants. Take care not to discolor or otherwise damage home furnishings when you spray. Apply only at recommended concentrations and when plant is not in its rest period.

FORCING. Forcing ensures a supply of fruits, flowers, or vegetables earlier than they would be available if cultivated in the usual way. Forcing necessitates the use of warmth, ordinarily supplied by hot-water or steam pipes, electric heating, or by a hotbed made of fresh manure and leaves.

Pussy willow (*Salix*) is a favorite for spring forcing. Cut the ends of the branches in January or February, place in water, and watch them unfold their large, closely packed catkins. Black pussy willow (*S. melanostachys*) is unusual and truly different; its catkins are so dark they appear almost black against the red twigs. (Wayside Gardens)

FORGET-ME-NOT (*Anchusa capensis*). These lovely, dainty plants are most effective in the garden when planted in large drifts. The popular variety Blue Bird is a compact, attractive plant with the bluest of all blue flowers. The flowers are too fragile for indoor arrangements and do not last well when cut. If cut as they fade, the flowers will bloom over a long period of time. The variety alba has white flowers.

Botanists say the hairy stems of many forget-me-nots are intended to keep ants and similar insects from stealing the nectar reserved for flying insects that pollinate.

FORSYTHIA (*Forsythia*) GOLDEN BELLS. The yellow blossoms of this lavishly beautiful shrub are one of the joys of February. Forsythias are outstanding as specimen plants and excellent for forcing. Branches cut in January and February will force in just a few days. Prune older wood immediately after blossoming to keep the shrub in good health and heavy flowering.

When food is scarce, birds may pick and tear at the unopened buds, but happily the plant has a reserve set which are rapidly brought into action if the season's normal quota is pillaged. Almond trees make good neighbors.

FOTHERGILLA GARDENI.   This early-blooming deciduous shrub is noted for its one-inch spikes of honey-scented, cream white flowers appearing in early spring. An outstanding characteristic of this shrub is the change of its leathery, dark green summer foliage to a spectacular display of brilliant yellow and orange red in the fall. To achieve this, plant shrub in full sun and in an acid soil with good drainage.

FOUR O'CLOCKS FOR FATAL FOLIAGE.   Japanese beetles, a pest on peach trees, like the foliage of the four o'clocks and are apparently unaware that in eating it they are committing suicide. Dig up the four o'clocks in the fall, and if carefully overwintered, they can be used again the following year. The foliage of this plant IS poisonous; don't let children or yourself take a bite of it.

FOXGLOVE (*Digitalis*).   Foxglove has a growth-stimulating effect on nearby plants and is a good companion for pines. It does well on a forest border and open woodland for naturalizing. Foxglove tea in the vase is said to prolong the life of other cut flowers.

The flower is the most valuable source of the powerful heart stimulant, digitalis. North American Indians knew and used foxglove for heart problems before it was known to Europeans. In England the extract was recommended by so-called witches (or herb women) for this purpose before it was recognized by physicians.

But be cautious if you grow this one in your garden. The tubular purple or white lavender flowers grow about two inches long in a twelve- to twenty-four-inch, one-sided cluster. The summer-blooming flowers are often spotted within. Severe poisoning comes from eating the fresh or dried leaves which do not lose their toxicity by cooking. Children have been poisoned by sucking the flowers and swallowing their seeds.

FRAGRANCE.   Flowers exude a powerful, seductive odor when ready for mating. This causes a multitude of bees, birds, and butterflies to join in a Saturnalian rite of fecundation. Unfertilized flowers emit a strong fragrance for as many as eight days or until the flower withers and falls; yet once impregnated, the flower ceases to exude its fragrance in less than half an hour. One tropical plant (*Colcasia odorata*) at time of flowering increases in temperature, repeating this phenomenon for six days from three to six each afternoon.

Fragrant flowers are usually light in color or white, with the purples and mauves coming next. Thick-textured flowers such as magnolias and gardenias are often heavily scented. The perfume of a plant is not always found in its flowers. It may be in the root, seeds, bark, the gum or oils, even in the leaves or stalk. Certain families, such as the Labiatae (lavender, rosemary, mints), are especially gifted with perfume.

FRANGIPANI *(Plumeria)* PERFUME TREE. There are more than forty species of these warm-weather ornamentals. With their exceedingly sweet smelling flowers, they are considered the most fragrant of ornamental plants. Their waxy blooms of deep rose and white consist of five petals overlapping in star fashion to a narrow throat supported by a thick short stem. Blooms form bouquets in clusters often eight to eleven inches across and continue opening in the same cluster for many weeks. In some parts of the country they bloom year-round.

Frangipani cuttings root easily, and if desired, the tree can also be propagated by seeds from the occasional paired, tightly filled, seed pods. Sizable trees are sold by nurseries to furnish immediate beauty for outdoor gardens and, where they are not hardy, they make lovely pot plants.

FROGS *(Amphibia)*. Toads and frogs are avid consumers of garden pests. It is estimated that a toad eats up to 10,000 insects in about three months and about 16 percent of these may be cutworms. It also gobbles grubs, crickets, rose chafers, rose beetles, squash bugs, caterpillars, ants, tent caterpillars, armyworms, chinch bugs, gypsy moth caterpillars, sow bugs, potato beetles, moths, flies, mosquitoes, slugs, and sometimes even moles.

In spring gather frogs for your garden from around the edges of ponds and swamps. Pen them in for a few days, otherwise their homing instinct may cause them to leave. Give them some shelter such as a clay flowerpot turned upside down with a small hole broken out of the side. Bury the pot in a shady place several inches in the ground. And don't forget to give them a shallow pan of water.

FROST PROTECTION. Evergreen branches, which are springy, are excellent frost protection for perennials and herbs.

FRUIT TREE MOTH. This moth is repelled by southernwood.

FRUIT TREES. Few things are lovelier than the blossoming of fruit trees in the spring. The crab apple is one such and is easily grown; its apples make a delicious jelly. Other attractive flowering fruit trees include dolgo, red dolgo, hyslop, and hopa. Fiery crimson crab is grown for its gorgeous blossoms but is laden in the fall with small scarlet crab apples that cling for a long time after the leaves have fallen.

Another valued ornamental is the Bradford pear which is one of the earliest trees to bloom in the spring. The abundant blossoms appear in clusters of ten or twelve. They are off-white, non-fragrant, and borne on

How to Plant a Fruit Tree.

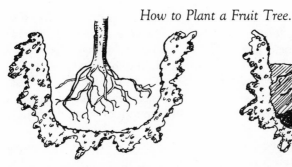

1. Dig a hole large enough so there's a few inches of space beyond the tip of the roots.

2. Place a mound of the topsoil in the center of the hole, then spread the roots out naturally.

3. Tamp down soil by walking on it. Add more topsoil until the hole is at least two-thirds full.

4. Pour in a bucket of tepid water. Let the water sink in gradually, then tamp again.

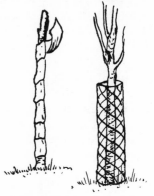

5. As you put in more soil, form a slight circular depression around the tree trunk.

6. Wrap the lower trunk with heavy paper or a cylinder of screen wire for protection.

short spurs. Collectively, they appear as a solid mass of white in vivid contrast to other spring foliage and flowers. The glossy green, thick, and broadly oval leaves appear just as the flowers start falling. Their wavy margins cause them to flutter in the wind and, like the flowers, they are abundant. The Bradford pear is gorgeous again in the fall. Early frosts bring about changes in the color of the leaves to deep hues of purplish red, then crimson. The tree rarely fruits, and when it does so, the fruit is inedible.

Dwarf fruit trees, grown espalier against a wall, provide an interesting and attractive way to use a narrow space.

Dandelions in the area stimulate fruit trees to ripen quickly. Garlic and onions contain powerful anti-bacterial agents and are an effective destroyer of diseases that damage stone fruits. Used as a spray three times daily at five-day intervals, a water solution of onion skin gives an almost complete kill of hemiptera, a parasite which attacks more than a hundred different species of plants. Myristicin, a compound in parsnips which has been synthesized, is a selective insecticide for fruit flies.

Other good companion plants are garlic (for borers), chives, onions, nasturtiums, horseradish, southernwood, and stinging nettle. A 15 percent interplanting of mustard with clovers is also helpful.

FUCHSIA *(F. exoniensis, F. corallina).* This vigorous, nearly prostrate shrub spreads to three feet or more. Its branches, covered with large, dark green leaves, arch gracefully. All summer long there is a sparkling display of long flowers with brilliant red calyces and a rich purple skirt. The shrub is beautiful for banks or to overhang a wall, and grows in sun or shade.

FUNGUS DISEASES. Plants are attacked by fungus, rusts, and other diseases for a variety of reasons. Extreme weather conditions, either drought or excessive rains, weaken plants. Peony leaves turn brown around the edges in wet weather. Or insect pests may injure them so that fungus diseases and rots find an opening. Other insects infest plants with certain virus diseases.

Raw organic materials used as fertilizer can cause a plant to succumb to whatever comes along because it cannot digest the crude material provided. The plant's chemistry is altered in such a way that it actually may attract insect pests. Well-decomposed compost is as good for your flowers as it is for your vegetables.

FUNKIA *(Hosta)* PLANTAIN LILY. This plant is characterized by its bold, colorful, fragrant flowers that are excellent for cutting and its leaves that are dramatically splendid in cut arrangements. *Hosta crispula* of Japanese origin has two-inch flowers of funnel form and lavender color, and each leaf has a serrated look. Protect from strong winds.

G

GARDENERS LIVE LONGER! A recent television program interviewed people who were well past 100 years of age. The common denominator found in each instance was their ability to cope with stress. The National Garden Bureau, Inc. states that gardeners live longer because gardening gives relief from tension, fears, and worries. High-tension people such as doctors, pilots, police, mothers of small children, and teachers escape to their gardens whenever possible. Stresses disappear in the familiar tasks of preparing soil, planting, cultivating, and harvesting. The garden is a place of healing, not just a "factory" for producing food or flowers.

GARDENIA (*Gardenia jasminoides*) CAPE JASMINE. This is a beautiful broad-leaved, evergreen shrub, two to six feet tall, with dark lustrous leaves and exquisite large, white, waxen flowers of enchanting fragrance. The double-flowered form is famous as a buttonhole flower. The shrub blooms from May to September in the South where it is often used for hedges. Use as a background for lower-growing flowers as the white blossoms blend well with all colors.

GARDENS IN A BAG. Improbable as it may seem, many people without backyards are starting to garden in bags, plastic bags filled with soil mix. Jiffy-Mix from Carefree Garden Products comes in three sizes (four, eight, or sixteen quarts of lightweight soil mix) in already packaged growing containers. To plant, cut crosses on top of the bag where you want to place plants. Two hours before planting, moisten soil thoroughly. Punch small holes in sides of bag for drainage. Add fertilizer according to manufacturer's directions.

A four-quart bag will hold two pepper plants or six small marigolds. Compact plants such as Patio tomatoes will grow well with no support. The new Explorer potato introduced by Pan American Seed Company also grows easily in these containers. Plant six plants in two groups of three at each end of the sixteen-quart size.

GARLIC (*Allium sativus*). Garlic is easy to raise, easy to store, and useful in the garden as well as the kitchen. Planted near roses it aids in fighting black spot. More commonly the garlic cloves are minced or placed in a blender, then mixed with water for a spray. Some add a tablespoon of vegetable oil to the spray to make it cling to flowers and foliage. The spray discourages many insects and combats various blights found on vegetables and flowers. The spray's antibiotic power is credited for this.

Garlic can be purchased in processed form completely deodorized. According to a scalp specialist, rubbing garlic on a balding head will restore hair.

Garlic cloves, chopped and taken internally, benefit the complexion. The fumes alone of freshly crushed garlic are so powerful they can kill certain disease-producing germs. Garlic is also of benefit for lung ailments in animals.

GENTIAN *(Gentiana).*   Ordinarily the gentians are without fragrance, but the rare perennial, fringed gentian has a delightful scent suggestive of strawberries.

Septemfida (var. *lagodechiana*) is an enchanting summer, flowering gentian with deep, true blue, one-inch blooms with a white throat. The dense, heart-shaped foliage, which likes summer heat, forms an attractive ground cover.

To sprout gentian seed, get a box about six inches deep, put in two inches of coarse gravel or broken crock pieces, then three inches of fine, porous soil. Plant the seeds in the soil and cover with one inch of peat moss. Sink the box in the garden in filtered shade and keep it moist until the seeds germinate. Treat other hard-to-sprout seeds like cyclamen, smilax, violets, cannas, Christmas roses, and nasturtium this way as well.

GERANIUM *(Pelargonium).*   The hybrid geraniums in scarlet, cherry, salmon, coral, and white are simply gorgeous. In addition to these new introductions, geraniums which could only be propagated by cuttings can now be grown from seed. Sprinter geraniums are free flowering and of unsurpassed garden performance.

Geraniums make good companions for roses because they repel Japanese beetles. Geraniums also repel cabbage worms when planted among the brassicas, and the white geranium is helpful when planted near corn.

GERANIUM *(Pelargonium)* SCENTED LEAVED.   In competition with the gorgeous Zonals, the sweet-leaved geraniums fell out of fashion, but the wheel is turning again. The sweet-leaved geraniums offer the best of scents and immense variety including the filbert, nutmeg, cinnamon, almond, lemon, orange, apple, anise, rose, pine, musk, violet, lavender, balm-scented, and many more. All of these also bloom but the flowers are inconspicuous.

A leaf of the rose-scented geranium in the bottom of the glass imparts a delicious flavor when making apple jelly. Use the oil of geranium as an insecticide against red spiders and cotton aphids.

GIBBERELLIC ACID.   In the mid-twenties a Japanese scientist working on rice diseases discovered a remarkable, colorless substance named gibberellic acid. Less than one drop of it on a camellia bud causes the flower not only to be much larger but also to bloom much earlier than usual. This has greatly extended the season of blossoming for camellias.

Another experiment has shown that a lawn can be made to start growing in early spring at below normal temperatures after being treated with gibberellic acid at a concentration of one ounce per acre. Scientists believe the gibberellins are natural plant products.

GINSENG *(Panax quinquefolium).*   Ginseng is the most expensive botanical in the entire vegetable kingdom and surpasses even the truffle as a precious aphrodisiac. Ginseng has been known and prized in the Orient for centuries, but it has not been highly thought of by occidental physicians. Recently, however, clinical and biochemical studies have found that ginseng has a beneficial estrogenlike effect on women.

Ginseng also stimulates the nervous system, making one feel more active, more aggressive, more interested.

GLADIOLUS *(Gladiolus).*   Thanks to Dr. Forman McLean of the New York Botanical Gardens, we now have sweetglads—scented gladiolus. Gladiolus are gorgeous flowers—tall and dramatically beautiful, coming in many brilliant colors. But keep them away from peas and beans, which dislike their presence.

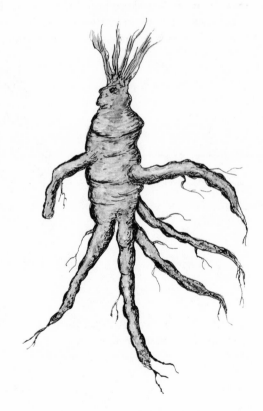

For centuries, Orientals have believed ginseng promoted long life and virility, and have used it as a cure for many ills. Americans who use the root praise it for its tranquilizing qualities.

GLOBE ARTICHOKE.   This elegant perennial dates back to the 16th century and is a form of cardoon, a favorite of the ancients. It prefers cool, often foggy summers, and is grown mainly in California. If allowed to blossom, the great thistlelike flowers are gorgeous. The edible buds (future flowers when allowed to mature) are a great delicacy and one of our most expensive "vegetables."

GLOXINIA *(Gesneriaceae).*   One of the most popular gift plants, this native of tropical America is remarkable for its richly-colored, velvety leaves and large, bell-shaped flowers. It is widely cultivated as a houseplant and in greenhouses.

GOLDENROD, SWEET *(Solidago odora).*   Anise-scented goldenrod is a delightful beverage plant. Collect the leaves when the plant is in bloom and use them fresh or dried, with peppermint leaves, as an after-dinner summer tea.

Goldenrods are also medicinal, for not only did the American Indian employ this common plant as a cure for sore throat and for pain in general, but today it is recommended by herbalists as a diaphoretic in colds and coughs, and as an aid in rheumatism. In fact, the generic name, *Solidago*, means, "I make whole," that is, heal.

Goldenrod is incorrectly blamed for hay fever and similar allergies; ragweed, which blooms at the same time, is the real culprit.

There are more than 100 species of goldenrod. It's used for dyeing, giving colors from mustard to brown olive, depending on the strength used.

GOLDENSEAL. ORANGEROOT, YELLOW PUCCOON, INDIAN TURMERIC. This is a native American herb. Its value was learned from the Indians by early trappers, hunters, and adventurers. Goldenseal grows in small colonies in rich woods. The greenish white flowers are ¼ inch across, followed by red fruits in large clusters.

GOLDFISH. If you have a pool in your garden, stock it with fish. Goldfish are excellent for consuming mosquito larvae, and so are the small topminnows called mosquito fish, *Gambusia affinis*.

GOURDS *(Langenaria)*. In times past every household grew a great variety of these versatile plants for many purposes. "Sugar trough gourds," according to the 1887 Burpee catalog, "are useful for many household purposes such as buckets, baskets, nest-boxes, soap and salt dishes, and for storing the winter's lard. They grow to hold from four to ten gallons each, have thick, hard shells, very light but durable, having been kept in use as long as ten years."

Dippers and birdhouses were made from the dipper gourd, spoons from the spoon gourd, or, left whole, it formed a darning egg for stockings or a rattle for the baby when the seeds dried inside. There was a further advantage—spoons and dippers stayed cool when stirring hot liquids.

After a few gourds have set, pinch off the ends of the vines and the gourds will grow larger. Harvest fruits when fully ripe and before frost; handle carefully to avoid bruising. With a sharp knife cut away unwanted parts to form dipper or container. Remove seeds, wash with a strong household disinfectant, and put to dry in a well-ventilated place. Turn every few days. This drying period may take several weeks or even months. Coating is not necessary for preservation, but some people polish their gourds with paste wax or coat with clear shellac when completely dried.

Buffalo gourd *(C. foetidissima)*. The powdered root is somewhat toxic to cucumber beetles; mix a little water with powdered soap for spray.

GRAPES. The sunlight on the leaves and not on the grapes determines whether or not grapes can be grown. Grapes will color normally if they have adequate leaf surface in proportion to the amount of fruit being produced. To obtain perfect bunches of grapes they are sometimes placed in bags. If you want to do this, use brown paper bags. Grapevines need a sunny, well-ventilated location. A grape arbor is so decorative that you may wish to place one in your flower garden. Hyssop planted near grapevines will increase yields.

GRASS. To keep neat edges on grass along flower beds, paint grass edges with tractor fuel. Use an old four- or five-inch paintbrush. This application will check Bermuda, carpet-, and St. Augustine grasses.

Though grapes are usually propagated by cuttings, they do have flowers, beloved of bees, and fertile seeds. Wild mustard is beneficial to grape vines.

"GRASS" FOR CATS.   Cats who rarely have the opportunity to go outside enjoy greenery to nibble on. Oats fulfill this need, providing a safe, healthy distraction from nibbling on houseplants, some of which may be poisonous. Catnip, a perennial mint, is also a favorite of felines, who are stimulated by its leaves.

GRASSHOPPERS.   The biological means of controlling grasshoppers, which are becoming an ever-increasing problem, is nosema grasshopper spore. *Nosema locustae* is a protozoan which specifically attacks grass-hoppers and some species of crickets. Dissolve it in water and add to a bran mixture, then disperse over the garden or yard area. As the grass-hoppers feed, they become infected and slowly die. Research has shown a 50 percent drop in population in a month. Furthermore, nosema is passed on from one generation to the next through the egg mass and is also transmitted when infected grasshoppers are eaten by healthy ones. (Plants of the Southwest, Park's Seeds)

GREENSAND.   Greensand is found along the New Jersey and Virginia coasts, among other places. This granular, half-soft marine deposit contains about 6 percent potash, and is an olive green, iron-potassium sili-cate, also called glauconite. Greensand is available from several natural-fertilizer companies. The Sudbury people recommend it as a major nat-ural source of potash as well as nitrogen and phosphorus. (See Sources of Supply.)

GREENS FOR SALAD. WILD CUCUMBER (ST. MICHAEL'S), BROOK SAXIFRAGE *(Saxifraga punctata L.).*   The flowering stalk grows four to twenty inches high, is hairy and leafless. The flowers are small, each with five white or purplish petals, in headlike or flattened open clusters at top of stem. The plant is found in moist, rocky, shady places along rivulets, roadsides, rocky cliffs, and gulches throughout southeastern Alaska, the Gulf Coast, and westward on the Alaskan Peninsula. Leaves for salad are collected in spring before the plant flowers and are a good source of vitamin C.

GROCERY PLANTS.   Check your grocery bag for some good winter growing projects. While such plants started from seeds cannot be depended upon to produce "true" fruits, or any at all for that matter, they are fun to grow. Lemons, oranges, grapefruit, and other citrus fruits are good for a start, and sweet potatoes make an attractive, interesting-to-grow vine.

GROUNDSEL *(Senecio vulgaris)* COMPOSITAE.   This common plant of waste places and pastures is of erect growth with grayish green leaves, branching with jagged lobes. Flower heads are in close terminal clusters, the individual flowers being of tubular shape, solid, and yellow, like minute candles, and possessing no ray petals. This herb is particularly rich

Grasshoppers damage crops and are of no value in pollinating plants. Not so with bees, without whom many plants could not reproduce.

Groundsel, or ragwort, is a favorite of caged birds who seem to understand the value of this herb, so rich in minerals.

in minerals, especially iron. Animals seek it out as a tonic, especially poultry, and it is relished by caged birds kept as pets. The herb has powerful drawing and antiseptic properties. Mixed with ground ivy it makes an important poultice. It strengthens the eyes and reduces inflammation.

GUELDER-ROSE *(Viburnum opulus)* Highbush Cranberry.   The highbush cranberry is a valuable wild plant yielding food, drink, medicine, and beauty, but it is not a cranberry. Nor is the almost identical Guelder-rose of England a rose. These tall shrubs reaching from six to ten feet are related to the honeysuckle, the elderberry, and the blackhaw. The attractive white flowers appear in showy clusters three to four inches across, with large sterile blossoms about the edge of the cluster and much smaller fertile ones near its center. The flowers are followed by bountiful clusters of bright red berries which become better tasting and soft when touched by frost. The berries hang on the bushes all winter. Birds eat them, but not until early spring when other food is scarce. This is an excellent plant for bird lovers to place in a wild garden.

GUNNERA INSIGNIA (Haloragidaceae or Gunneraceae). These ornamental perennials are mostly from the Southern Hemisphere. They prefer to live on the inaccessible mountainsides of Guatemala and Panama. The huge leaves measure from four to five feet across and resemble our common pot geranium. Gunnera is of such massive proportions that usually it is only planted in public grounds for landscape effects.

One leaf is capable of serving several people as an umbrella. Gunnera bears a large stalk covered with thousands of small, brownish, wind-pollinated flowers. A similar species of gunnera is found in the mountains of Hawaii.

H

HANDICAPPED. Gardening is wonderful therapy for the physically and mentally impaired. Children of all ages, including retarded and underprivileged children, love to "make things grow." And wheelchair gardening can add interest to a life that may often be dull and monotonous. Even a box or raised bed of suitable height in which to grow flowers or vegetables (or both) can brighten the hours as seeds sprout, leaves unfold, and flowers bloom.

HANG IT ALL! Hanging baskets never fail to attract attention, be it cascading petunias gracing an old-time porch, a gay ivy geranium attached to a lamp post, or a dainty fuchsia suspended from a tree branch. Hanging baskets have a magic all their own.

A hanging basket is really a pot plant gone glamorous and needs to be fed to keep going. Jobe's Flower Spikes are a handy way to do this.

Here are some suggestions of suitable candidates for various positions.

Plants to grow in sun.
Flowering: lantana, ivy geranium, phlox (annual), lobelia, dwarf French marigold, nasturtium, oxalis, petunia, bougainvillea, sweet alyssum, verbena, monkeyflower (*mimulus*), shrimp plant, dimorphotheca, pinks (*dianthus*), cascade chrysanthemums, pansy (early spring), Tiny Tim tomato.
Foliage: English ivy, Sprenger asparagus, donkey-tail (*Sedum morganianum*), variegated vinca, peppermint geranium, purple passion (*Gynura sarmentosa*), flowering inch-plant (*Tradescantia blossfeldiana*), setcreasea, siebold sedum.

Plants to grow in shade.
Flowering: fuchsia, achimenes, browalia, coleus (trailing "Queen"), black-eyed Susan vine (Thunbergia), tuberous begonia, columnea, episcia, flowering maple, star of Bethlehem (*Campanula isophylla*), torenia.
Foliage: English ivy, German ivy, Swedish ivy, grape ivy, kangaroo

ivy, pick-a-back plant, variegated arch-angel, spider plant (Chlorophy-tum), inch plant, zebrina *(tradescantia)*, philodendron, rosary vine, Christmas cactus, strawberry "begonia" *(saxifrage)*, epiphyllums, pa-tience plant, pothos, Boston fern, rabbit's foot fern (and other ferns).

HAWTHORNE *(Crataegus oxyacantha)*.   The hawthorne is a beautiful May-blooming shrub with sweet-scented blossoms and lovely pink or rose or white double flowers. Paul's scarlet thorn is not fragrant. And the blossoms of some of the American hawthornes have a disagreeable odor. Be sure of the kind you plant.

Hawthornes make excellent hedges around the flower garden and for windbreaks, shade in hot weather, and protection against intruders.

HAY FEVER SUFFERERS.   Knowing what and when to avoid is help-ful. The eucalyptus tree spreads its misery from January through Sep-tember. March, April, and May are particularly rough periods for those allergic to sycamore, English walnut, and live oak. Springtime also brings the cottonwood fuzz. Bermuda grass can keep you down from March through November. April and May are the toughest grass months with pollen from fescues, rye, and Kentucky blue keeping your eyes watering. Johnson grass hits later from May through August.

Weeds of all types pollute the air, starting in May and running through October. Ragweed's worst two months are August and Septem-ber—and don't blame the innocent goldenrod.

Besides flowers, trees, weeds, and grass there are some common year-round environmental allergy-producers such as: cat hair, cattle hair, chalk, dog hair, glue made from animals, horse hair, house dust, news-paper, sheep wool, tobacco, and feathers. Foods that are most often allergy-producers are: wheat, celery, chicken, cow's milk, chocolate, eggs, oranges, peanuts, strawberries, tomatoes, and cantaloupe.

HEADACHE.   The Romanies have a number of interesting flower and herb remedies:

(1) A few pieces of willow bark boiled in a pan of water is helpful. It is their salicylic acid, the ingredient found in most patent medicines.

(2) For nervous headaches the flowery tops of rosemary made into a tea with boiling water is soothing.

(3) A tea made from a few dried lime flowers cures a headache in about half an hour. Take hot, then lie down for half an hour and relax.

(4) For a severe headache put a pinch of dried marjoram into a tea-cup. Half fill it with boiling water, cover and allow to draw, and drink while hot.

HEATHERS AND HEATHS (ERICAEAE).   These dwarf evergreen shrubs are excellent for edging or in front of taller evergreens in a foundation planting. Their foliage, which persists in winter, takes on attractive shades of green, bronze, and gold, adding winter interest to plantings which contain them.

Heathers and heaths resemble each other closely in their growth characteristics, however, the heathers (*Callunas*) are hardier than the heaths (*Ericas*). Heathers flower for the most part in summer and fall, some continuing into winter, while the heaths bloom in late winter and spring. They come in shades of deep rosy red, brilliant pink, and white. Heathers may be lilac mauve, silvery pink, red purple, and pure white. The heath *Erica carnea*, Springwood White, is one of the first shrubs to bloom in winter.

HEN-AND-CHICKENS (*Semperivum tectorum*) HOUSELEEKS. These plants grow without care in a sunny spot. There are sixty or more varieties from tiny green cobwebby ones to sturdy copper colored ones like Heuffell. They need little soil and are fine for covering walls, rocky hillsides, or small pockets in rocks or paving. They are hardy and colorful in late winter. The Romans thought they would protect a house from lightning if they grew on the roof.

HEPATICA ACUTILOBA. These perennials bloom in early spring. Their pink, softly colored flowers resemble anemones and are surrounded by sharply pointed leaves which often persist through winter. Just six inches tall, these wildflowers are fine for naturalizing.

HERBAL REMEDIES. The information given below is not intended to replace a physician, but there are times and circumstances when a doctor is not available, and minor distresses plague us. In choosing herbal remedies, some simple definitions may be helpful:

Alteratives improve nutritive processes gradually, normally, and naturally. Take in the form of tea. Peppermint tea is a for instance; it is good for relieving colic in babies or minor bloat in children and adults.

Antispasmodics reduce involuntary contractions often arising from nervous causes. *Anthemis nobilis*, or true camomile, is notable for giving relief. Camomile flower tea has been famous since the days of the Pharoahs and was as popular in past times as aspirins are today.

Carminatives and Aromatics are usually herbs with a spicy scent and pungent taste useful for expelling gas from the stomach and thereby reducing flatulence. Chew mint, a mild carminative, after dinner. Other herbs which may be used separately or together are anise and caraway, cloves, dill, ginger, and other aromatic spices. Their delicious fragrances also provide a lift to our spirits.

Demulcents sooth the intestinal tract and are usually of an oily or mucilaginous nature. Olive oil is a natural demulcent. *Emollients*, similar in meaning, are used to soothe the skin rather than internal membranes, often allaying the pain of irritated parts. Irish moss and slippery elm are both demulcent and emollient.

Diaphoretics produce sweating. Sweat baths have been used by many races over the centuries. In the Finnish sauna, water is thrown on heated stones, accompanied by stoking with cedar or birch boughs. Yarrow tea (*Achillea millefolium*) induces sweating and formerly was used as a cold remedy. Dried elder flowers and mint are helpful added to the brew.

Diuretics stimulate the secreting cells or nerves of the kidneys, increasing the flow of urine. Herbs with diuretic value are dandelions, juniper berries, and lemon juice; milk is also a mild diuretic.

Expectorants help in the expulsion of mucous secretions from the air passages, such as phlegm which accumulates in the lungs and windpipe. Yerba santa, red clover tea, and valerian are examples; there are many more herbs which are also helpful.

Febrifuges are agents that reduce fevers. Dogwood and boneset, sometimes called feverwort, were used for this purpose.

Laxatives are safe to take to relieve a temporary condition of constipation, but plants that are *cathartic* in action should be carefully considered. Dandelion greens are a tonic and mildly laxative.

Nervines act upon the nervous system to overcome irritability. The most notable plant for this purpose is valerian.

Sedatives help to quiet the nervous system without producing narcotic effects. The bark of the wild black cherry is added to cough medicines because of its sedative action. Rue is also used as a tonic with sedative qualities.

Stimulants quicken vital action and digestion, raise body temperature, and increase general awareness. Culinary herb stimulants are anise, pepper, cinnamon, cloves, dill, ginger, horseradish, nutmeg, peppermint, and sage. Medicinal herbs include horehound, hyssop, lavender, lobelia, marjoram, rue, and spearmint.

Stomachics improve stomach activity and stimulate appetite. Spearmint is an excellent example. Chokecherry, often used for making brandy, is another.

Tonics tone up our bodies and give us a feeling of well-being. They are often referred to as "bitters." Camomile is such a one, so are feverwort, chicory, goldenseal, *Verbena officinalis*, and wild tansy.

HERB OF GRACE (*Ruta graveolens*) Rue.   This is a very bitter tasting herb, strongly aromatic, and once important for medicinal purposes. Plant rue with roses to foil the Japanese beetle. It is also helpful grown with fig trees. Cats detest rue so rub it on your upholstered furniture to keep them from clawing. Use against fleas in the dog's bed, and a few sprigs hung in a room will drive out flies. Rue and basil are incompatible.

Sage, one of the easiest herbs to grow, is commonly listed both for cooking and for medical uses. Its name comes from the Latin salveo, *meaning to save or heal.*

HERBS, BENIGN. Some plants down through the centuries were thought to have a benign effect while others brought evil in their wake. The "good" plants were southernwood, rosemary (effective against witchcraft), lavender (against the evil eye), bracken, ground ivy, maidenhair fern, dill, hyssop, agrimony, and angelica. Yellow and green flowers growing in hedgerows were believed to be especially disliked by witches.

HERBS, EVIL. Other herbs had the opposite effect and were the means of invoking evil spirits. The herbs to call up evil spirits were vervain, betony, yarrow, and mugwort. St. John's-wort was used to exorcise them. How herbs were used made a lot of difference, also the combinations. It was believed that if coriander, parsley, hemlock, liquid of black poppy, fennel, sandalwood, and henbane were laid in a heap and burned together, they would call forth a whole army of demons.

HERBS, STREWING. In northern countries, before rugs were commonly used, the floors of castles and churches were strewn for warmth with various organic materials such as rushes. Herbs, called strewing herbs, were popular, and lavender, thyme, *Acorus calamus*, the mints, basils, balm, hyssop, and santolina were widely used. Marjoram, believed to be an antiseptic, was scattered over church floors at funerals.

HIGH BLOOD PRESSURE. Recommended fruits and vegetables help-ful in reducing high blood pressure are: broccoli, carrots, cauliflower, celery, cherries, cranberries, cucumbers, desert tea (Ephedra), endive, fennel, garlic, grapefruit, guavas, kumquats, melons, oranges, peaches, pears, peppers, pineapple, pomegranates, raspberries, spinach, strawber-ries, tangerines, and turnip tops.

HOARSENESS. This is another Romany remedy. Take a good-sized turnip, wash it well but do not peel. Then cut a piece from the bottom so that it will stand upright, and cut it downward in four equal slices. Fit the turnip together again to stand up in a deep dish or soup plate, hav-ing first added a layer of demerara sugar or of honey between the slices.

When the turnip has been left standing for an hour or two, a thick syrup from the turnip juice and the sugar or honey will have formed in the bottom of the dish. This can be taken a spoonful at a time.

HOLLYHOCK (*Althaea rosea*). This native of China is a tall perennial herb which is usually treated as a biennial in our gardens. The roots are demulcent and emollient, thus of good use as cold and diuretic reme-dies. The generic name, *Althaea*, is derived from the Greek *Althainein*, meaning "to heal." Hollyhock is good to rub on bee stings if the leaves are bruised in oil and made into an ointment.

HOLY HERB (*Eriodictyon californicum*) YERBA SANTA. The small white and lilac flowers are fragrant. The leaves are strongly aromatic when crushed and were used medicinally by the Indians for colds, hence the name "Holy Herb."

HONEY. Honey drinks have been popular since time immemorial. Mead or metheglin, as it was called by the Welsh, contains the flowers of elder, rosemary, and marjoram, a handful each, with cloves, ginger, cinnamon, and pepper added to taste. The hydromel of the ancient Teutons was a honey wine drunk by them for thirty days following mar-riage—whence comes the expression "to spend the honeymoon." In Rus-sia a drink termed lipez is made from the delicious honey of linden. The mulsum of the ancient Romans consisted of honey, wine, and water.

Honey varies greatly in taste depending on the flowers the bees visit. This also affects the color. The most important honey plants in the United States are white clover, sweet clover, basswood, buckwheat, tu-pelo, raspberry, milkweed, goldenrod, alfalfa, fruit trees of all kinds, aca-cia, and mesquite.

HONEYSUCKLE (*Lonicera*). There are numerous kinds of honey-suckles characterized by the sweet honeysuckle scent and full of honey for the joy of the bees.

In fairly recent times a decoction of the stems was used for the gout, while an infusion of the flowers was believed helpful for asthma sufferers.

The winter honeysuckle (*L. fragrantissima*) suggests the scent of roses. Blooms usually occur before leaves open, and the fragrance of the profuse, tiny blossoms carries for yards from late February to April. The plant grows well in sun or shade even in a northern location, and sometimes is evergreen in a sheltered spot.

HOPS *(Humulus lupulus)* BEAR HOPS. Prepare the young sprouts like asparagus or use as a potherb for soup. At one time ground hops were a substitute for baking soda. The greenish yellow flowers are used commercially in the preparation of beers and ales.

A pillow of hops, instead of feathers, will induce sound, refreshing sleep. Powdered hop leaves are toxic to Southern armyworms and melonworms.

HOREHOUND *(Marrubium vulgare).* The wild and woolly horehound is not a very decorative herb; the flowers are small, mintlike, and white, while the leaves are covered with a dense, felted wool that gives the whole plant a downy, whitish appearance. Unlike most mints, the flavor and medicinal properties of horehound are not volatile or easily lost, so the plant can be used fresh or dried or boiled without driving off the flavor. Horehound tea is good for colds and horehound candy is good for coughs.

To make horehound candy, prepare a small decoction by boiling 2 ounces of the dried herb in 1½ pints of water for about 30 minutes. Strain this and add 3½ pounds of brown sugar. Boil over a hot fire until it reaches the requisite degree of hardness (testing from time to time in a cup of cold water). Then pour into flat tin trays, previously well greased. Mark into sticks or squares with a knife as it becomes cool enough to retain its shape.

HOUSEPLANTS. Here are some suggestions for preventing problems with houseplants:

Overfeeding. Don't fertilize houseplants too often. Once a month with a dilute liquid fertilizer is sufficient. Remember, don't fertilize a plant to make it grow, fertilize it because it is growing.

Overwatering. Underwater rather than overwater. More plants are killed by overwatering than by anything else. A dry surface appearance is a poor test of water needs because the drier atmosphere of a house quickly dries out the surface.

Poor Lighting. Proper lighting is difficult to achieve in city homes and apartments. The best way to overcome this is to grow most plants under the new full-spectrum fluorescent lights. Flowering houseplants

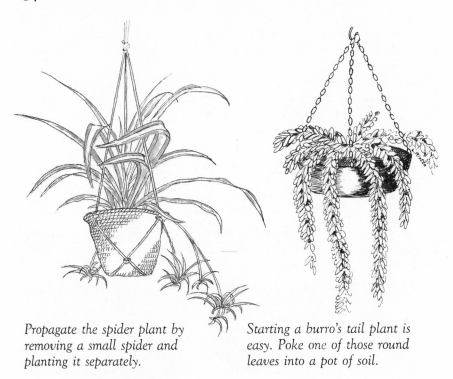

Propagate the spider plant by removing a small spider and planting it separately.

Starting a burro's tail plant is easy. Poke one of those round leaves into a pot of soil.

It's called dumb cane because if you bite its roots, you will be speechless for days.

The dracaena plant has leaves much like those found on a palm tree.

must have light, so reserve the sunfilled windows for them. Foliage plants prefer indirect light without sun. Only a few can get along in the dark hallways.

Improper Humidity. Proper humidity is very important. Most homes lack humidity in the winter due to radiators and heating units that dry the air. A humidity of between 50 and 60 percent is best for most plant growth.

Too Much Heat. Most homes are warmer than the recommended temperature for houseplants. A cool room is much preferred to a warm one for plant growth. Plants require air circulation even in cold weather, but avoid placing any plant in cold drafts or where air from central heating systems will blow directly on it.

Remember plants are not like animals or people. They cannot refuse the food given them, nor can they move to a more favorable location. See INDOOR PLANTS.

HUMMINGBIRD FLOWERS. In our western gardens alone there are fifteen different species of tiny iridescent hummingbirds, the ruby-throated being the only one widespread in the East. Hummingbirds have excellent color vision and are easily attracted by bright red flowers which they fly to from great distances. However, once in your garden they will visit flowers of any color in their search for nectar and small insects.

Without feathers, the smallest of the hummingbirds is no larger than a bumblebee. This bird's long, slender bill is especially suited for sucking nectar from deep-throated flowers such as honeysuckle and the trumpet flower.

Hummingbird flowers are long, tubular, contain copious nectar, and are often borne sideways or drooping rather than upright. Hovering before the flowers, the tiny creatures insert their long bills and tongues while whirring their wings more than 3,000 times a minute while they feast.

Good flower choices for hummingbirds also provide color all season and include: red columbine (*Aquilegia elegantual*); Indian paintbrush (*Castilleja integra*), blooming in the spring; scarlet bugler (*Penstemon barbatus*); skyrocket (*Ipomopsis aggreta*), summer; and hummingbird trumpet (*Zauschneria latifolia*) in the fall. Some species, such as Indian paintbrush, scarlet hedge-nettle (*Stachys coccinea*), and autumn sage (*Salvia greggii*) begin blooming in early spring and are stopped only by fall frosts. Other possibilities include desert beardtongue (*Penstemon pseudospectabilis*), shocking pink flower; scarlet runner bean, very showy scarlet flowers; scarlet larkspur, red orange flowers; toadflax, violet to magenta; Rocky Mountain columbine, blue and white; *Penstemon polmeri*, pink to salmon pink; and foxglove, purple flowers. (Plants of the Southwest)

HUNTING BOTANICALS.   The Indians imitated the scent or spoor of the deer with roots and herbs. The roots of blue wood aster and others were used to make smoke to attract the deer so that they could be shot with bow and arrow. Other plants used to attract deer were large leaved wild aster, root smoked; Canada fleabane, disk florets smoked; Philadelphia fleabane, disk florets smoked; and swamp persicaria, flowers smoked.

To sharpen their powers of observation, Indian hunters drank a tea made of heal-all root (*Prunella vulgaris*).

Indians of the Great Lakes used botanicals for trapping and fishing. The smell of the root of alternate-leaved dogwood, boiled in water, was used to disguise muskrat traps. Mountain mint was used as a lure on traps to catch minks. Other fur-bearing animals were attracted by a wash made from roots of kidney liverwort. Traps were boiled in maple bark to deodorize them so the animal would not detect the scent of previous victims.

The roots of wild sarsaparilla, mixed with roots of sweet-scented calamus, were boiled in water to make a lure for fish. Nets were then soaked in this brew; the scent would cling to the nets even after they were immersed for many hours.

HYACINTH (*Hyacinthus orientalis*).   These fragrant flowers are natives of eastern Mediterranean and western Asia. The plant has a large bulb with a purple or white scaly covering and from four to six linear-lanceolate, hooded, bright green leaves. The flowers are produced in spring on a terminal cylindrical cluster on a central stem from six to twelve inches high. In cultivated varieties the flowers are white or various shades of red and blue. The white and blue companion well with tulips by offsetting their flashing, brilliant colors.

HYBRIDS. The production of new flowering plants occurred only after the sexual basis of plant reproduction, along with the importance of heredity and genetics, was understood. It was then possible to take a plant with one desirable characteristic (for example flower size) and cross it with a plant having another desirable characteristic (such as a specific color). Hybrid plants are thus formed by taking the pollen from one plant and transferring it to the ovary of another. The resultant seeds are first generation hybrids and can be planted to produce the superior flowers, fruits, and vegetables found in seed catalogs. Hybrids also produce higher crop yields and disease resistant plants. See F_1 HYBRIDS, F_2.

HYDRANGEA *(Hydrangea macrophylla).* As many as thirty-five species of these deciduous shrubs or vines with large bold flower clusters and leaves are found growing in the United States. *H. macrophylla,* widely called hortensia, is the pot or tub hydrangea that florists force for spring bloom.

Potted hydrangeas are often made to produce blue flowers by adding aluminum sulphate to the soil weekly at the rate of ½ pound to 5 gallons of water. After two or three applications, apply 4 ounces of ferrous sulphate per 5 gallons of water for a few weeks. This treatment is continued as long as it is necessary to keep the soil acid.

Blue flowered hydrangeas produce flowers of various pinks if grown in non-acid soil; a neutral or slightly alkaline soil will give the results you desire. Add sufficient lime to raise the pH of the soil to a figure between 6.7 and 7.2. The addition of lime works best in the fall. Lift out the hydrangea. Shake the roots free of as much soil as safely possible. Then mix the lime thoroughly with the soil before replanting. Have your soil tested if you want to be certain.

H. macrophylla is one of several that is extremely poisonous. Cyanide compounds are present mostly in the leaves and branches.

HYSSOP *(Hyssopus officinalis).* This plant with its dark blue blossoms, neat linear leaves, and bushy growth is attractive in the flower garden. It may be clipped like boxwood to make a low hedge although this will be at the expense of the blossoms. There are also white and pink varieties. Hyssop is good to plant with grape vines. Planted with cabbage it will deter white cabbage butterflies.

I

ICE PLANT *(Mesembryanthemum crystallinum).* This one is a little trailer and creeper grown for its thick, succulent foliage and tiny, white blossoms. The fat, fleshy leaves are covered with glistening dots that have the appearance of ice, thus the popular name. The plant may be started from cuttings and does well in the driest and thinnest of soils. Try it in hanging baskets, window boxes, and rock gardens.

ICHNEUMONID WASPS. These are parasites of moth and butterfly larvae. The adult wasps feed on pollen and nectar, and often from puncture wounds made in host larvae.

IDENTIFYING SPECIMENS. To identify poisonous plants, contact your local Poison Control Center. New York State College of Agriculture offers a mail-order service to identify plants for people with no local means of assistance. Packaged plants may be sent to Extension Specialist, D.H. Bailey Hortorium, Ithaca, NY 14850, with an accompanying letter explaining exactly where the plant was found and the date it was collected.

IDESIA POLYCARPA *(Idesia).* This attractive deciduous tree is found wild in southern Japan and in central and western China. The Chinese type is hardy as far north as Boston, Massachusetts. The yellowish green flowers are followed by bunches of small fruits resembling bunches of grapes. Fruits are red when ripe. Some trees produce all female flowers, others all male, and yet others, both. The all-male-flowered trees do not fruit and the all-female-flowered ones only do so when a male tree is nearby.

IMMORTAL FLOWERS *(Xeranthemum annum)* IMMORTELLES. Sometimes winter flower arrangements are a problem. Of course there are pot plants, bowls of bulbs, and evergreen arrangements. But give thought to ordering a few packets of everlasting flowers.

Helichrysums, strawflowers, have been much improved in recent years and now include such colors as gold, red, rose, and salmon. Large sprays of these with fully opened flowers and buds may be dried in bunches, hanging down. A wire clothes hanger is handy for this.

Helipterums, sometimes sold under the names of rhodanthe and acroclinium, are among the prettiest of the immortelles and are easy to grow. Pick the flowers as soon as they are open and, again, dry head down.

Annual sea lavender, or statice, grows in a selection of shades. *Ammobium alatum,* a hardy annual, has silvery flower heads. *Catananche caerulea major* with lovely blue flowers must be picked and dried as soon as the flowers open or they will fade. *Xeranthemum annum* is double flowered and comes in white, purple, or mixed shades. Anaphalis has pearly white flowers and gray foliage. Do not pick until the flowers reach the papery stage. *Didiscus caeruleus* has many lavender blue flowers on long stalks. *Amaranthus caudatus*, or love-lies-bleeding, is very pretty. Physalis, the old Chinese lantern, and lunaria, or honesty, are excellent in mixed, dried bouquets. The yarrows (achilleas) and the silvery sprays of lamb's-ears (*Stachys lanata*), should also be dried upside down. Bells of Ireland (*Molucella laevis*) has wonderful eighteen-inch sprays of tiny flowers with large green calyces. It is as pretty and green in a summer arrangement as it is silvery and dried in winter.

INCENSE. Incense is predominantly of plant origin, a mixture of sweet-smelling gums and balsams. It burns with a delicate fragrance. The early Egyptians burned it at religious ceremonies; the Greeks and Romans and later the early Christians adopted this practice. The burning of incense is still part of the ritual of the Eastern Orthodox church, the Roman Catholic church, and some Episcopalian churches. Buddhists also burn incense at religious ceremonies. And it is often burned in homes to give fragrance to a room.

Incense ingredients of ancient civilization and the early Christians were: *Frankincense galbanum*, myrrh, mastic, rosemary, opopanax, and storax.

Incense ingredients of oriental nations were: cinnamon, cloves, camphor, dragons blood, galbanum, sandalwood, and star anise.

INDIAN BLANKET (*Gallardia*).   Our native gallardias have been developed into a variety of horticultural forms. The rich yellow and red daisies thrive in hot, dry places and bloom all summer long. They are fine as cut flowers and are easy to grow. Plant Indian blanket with calendula, Iceland poppies, and Maltese cross.

INDIAN LICORICE (*Arbrus precatorius*).   This plant is so keenly sensitive to all forms of electrical and magnetic influences that it is used as a weather plant. Botanists who first experimented with it in London's Kew Gardens found in it a means for predicting cyclones, hurricanes, tornadoes, earthquakes, and volcanic eruptions.

INDIAN PAINTBRUSH (*Castilleja*).   These perennials grow to 1½ feet tall. Their large, showy, brilliant red to orange, leaflike bracts surround the tiny, inconspicuous flowers.

Partial root parasites make this plant difficult to transplant, but it can be grown from seed if seed of another plant is sown in the same pot. Blue gramma (*Coutelous gracilis*), a grass, will encourage the seed to sprout. Indian paintbrush is thought to be parasitic on other plant roots.

Seed may be sown outdoors on bare ground around sage or clumps of

Indian paintbrush, which grows profusely in the Southwest, has a long period of bloom, and it's a favorite of humming-birds.

native grasses. The plant has a long period of bloom and is attractive to hummingbirds.

INDIAN PIPE *(Monotropa uniflora)*.   These strange parasitic ghost flow-ers never fail to shock when we unexpectedly come upon them, clus-tered in seemingly sinister closeness, deep in some somber woodland. At certain periods of their development the waxen white or flushed flowers give forth a "delicate and wholly sweet scent." They are often found in dry woods, usually under pine or beech trees, and get their food from decaying plants in the soil.

INDIAN RICE *(Fritillaria camchatcensis)* CHOCOLATE LILY, KAMCHATKA LILY.   This perennial plant has a simple stem, one to two feet, arising from bulbs with thick scales. The flowers, one to six, large, nodding, bell-like, dark wine color, are often almost black tinged with greenish yellow outside and have three petals. The bulb of large scales is sub-tended by numerous ricelike bulblets. The plant grows in open coastal meadows in southeastern Alaska, Gulf of Alaska coast, and north to Talkeetna, Alaska Peninsula, Kodiak Island, Aleutian Islands, and Bris-tol Bay area.

Bulbs are dug in the fall; they are then dried and used in fish and meat stews or pounded into flour. They are used extensively by natives of southeastern Kodiak and the Aleutians.

INDIAN TURNIP (*Arisaema dracontium*). The corm of the Indian turnip has insecticidal properties.

INDIGO, FALSE (*A. fruiticosa*). Acetone extract of the flowers is repellent to chinch bugs and striped cucumber beetles. Powdered mature pods with seeds are moderately toxic to Mexican bean beetle larvae. A sugar derivative, amorpha, is effective as a dust against chinch bugs, cotton aphids, squash bugs, tarnished plant bugs, potato leafhoppers, blister beetles, and spotted cucumber beetles. The fruit is more toxic than the roots.

INDIGO BUSH (*Amorpha canescens*). Indigo bush is a deciduous shrub with narrow spikes of lovely, tiny purple flowers in late summer. Because of a deep tap root, it is drought tolerant. Plant in full sun or partial shade. *Amorpha frucicosa* is similar but much larger with attractive flowers in large clusters, and butterflies adore it.

INDOOR PLANTS. Cosmetic care for indoor plants keeps them healthy. Remove dead or old flowers and yellowed leaves; these can harbor and encourage insect and disease problems. Wash foliage periodically. Dust and grime block out light and clog the stomata cells, which allow the transfer of gases from plants to the atmosphere. Syringing also helps reduce some insect populations such as spider mites and mealybugs, and the cooling effect stimulates growth.

However, be cautious of syringing espicias, gloxinias and most other gesneriads, and succulents when water is cold; spot damage will occur on leaves. Time your hosedown so water will not be on your plants at night, or fungal disease will result, especially during cool months. Collect and use rainwater if your tap water is high in iron, carbonates, or other dissolved minerals. See HOUSE PLANTS.

INFLORESCENCE. The word inflorescence means "a flowering" or a "flower cluster." The largest known inflorescence is that of *Puy raimondii*. The jolly green giant is a rare Bolivian plant with an erect panicle (diameter eight feet) which emerges to a height of thirty-five feet. Each of these bears up to 8,000 white blossoms. This is also said to be the slowest flowering plant, the panicle emerging only after 150 years of the plant's life. After blooming, it dies.

INSECTICIDAL FLOWERS. Asters, chrysanthemums, cosmos, coreopsis, nasturtiums, French and Mexican marigolds are good to plant throughout the flower garden.

INSECTICIDES. Here are three insecticides that are believed to be safe. Nicotine sulphate, commonly sold as Black Leaf 40, is a tobacco derivative that kills insects on contact. It is poisonous to humans but easily washed off.

Rotenone, a powder made from derris root, is used either as a dust or spray and will kill aphids and grasshoppers. It also kills fish so be careful using rotenone around water.

Pyrethrum powder, made from pyrethrum or painted daisy, is good against aphids and chewing insects. Pyrethrum is of little danger to man or animal, and insects do not develop resistance to it.

Follow manufacturer's directions carefully in using any of these preparations.

Another insecticide is Safer Agro-Chem's Insecticidal Soap to control aphids, mealybugs, whitefly, spider mites, scales, psyllids, earwigs, rose slug, crickets, and spittlebugs on houseplants, ornamentals, and in greenhouses. The liquid concentrate is mixed with water and applied as a spray.

INSECT STINGS, GUARDING AGAINST.   Problems arising from insect stings can range from mere annoyance to fatal. All sorts of stinging insects are encountered from time to time by gardeners, but the four most likely to offend are hornets, honeybees, yellow jackets, and wasps.

To avoid trouble, don't go around barefoot, and be careful mowing the lawn, cutting vines, or pulling weeds. Wear neutral-colored clothing and avoid floral prints when working around the garden or yard. Insects aren't terribly bright; a pretty patterned shirt or blouse can look to an insect like a bunch of flowers. Also attractive to insects are perfume and hair sprays. If they begin buzzing and you try to shoo them away, you will only aggravate them. So far there is no effective repellent against stinging insects.

Types of stinging insects vary from one geographical area to another; yellow jackets abound in the mid-Atlantic states, honeybees are found mostly in agricultural areas, and paper wasps are prevalent in the south central region and in parts of the southwestern states.

A new vaccine made from insect venom has been developed to protect people from life-threatening reactions to insect stings. There are also special emergency kits which can be purchased with a doctor's prescription. Included in the kit are chewable antihistamine tablets, a tourniquet, alcohol swabs, and a loaded syringe. If the injection is given as soon as symptoms appear, a severe reaction is unlikely.

If you or some family member are allergic to insect stings, knowing about these precautions may save a life.

INTENSIVE CARE.   An intensive care spot for plants can be a sheltered area under shrubs where houseplants can safely spend the summer, or possibly a place near porch or patio to start seedlings or root cuttings, or a cold frame for starting young seedling plants. Use these places for plants that have been on display in dark or smoke-filled rooms, in drafts, or exposed to too much sunshine. Plants can often be brought back to health and become attractive again if given a period of rest under more ideal conditions.

Also keep extra plants growing in a secluded area for replacement purposes as the life spans of earlier plants are reached and they must be pulled. Young zinnias make a fine summer replacement for more delicate spring plants which must eventually be removed. Keep some young shrubs or evergreens growing in case you lose one by death or disfiguration. Prepare for such misfortunes by starting extra specimens in your intensive care "unit" in case of need.

IRIS *(Iris)* FLEUR-DE-LIS. The name comes from the Greek word for "rainbow." These perennial plants in many attractive colors grow throughout the temperate region blooming in spring, summer, and a few in autumn. Many of the tall bearded varieties are gorgeous, earning for themselves the name "poor man's orchid."

Its worst enemy is the iris borer. If you find these, dig up the plants, clean out the borers, and dip the rhizomes (roots) in diluted Clorox (one part Clorox to ten of water) before replanting.

The dried root is called orris root. It is used in perfumes, powders, and medicines.

Iris, one of the earliest plants to bloom in the garden, teams well with grape hyacinth and daylilies.

IRISH MOSS *(Carrageen)*.   This is the name of several kinds of seaweed that grow in rocky places of Great Britain, Ireland, and the eastern coast of North America. The most common type has thick, forked fronds which are fan-shaped and colored reddish or purple. It is often used to make a nutritious jelly. The extract carrageen is used by confectionary and bakery firms to prepare low-acid content sweets. Manufacturers of chocolate milk drink find the extract valuable as a suspending agent. Mixed with cucumber juice, Irish moss offers a soothing hand lotion to apply to chafed or winter-chapped skin.

ISMENE *(Hymenocallis)*.   Ismene is a real treasure for southern gardens where the bulbs are hardy. In June four to five giant funnel-shaped flowers are borne on each stem. The flowers are richly perfumed and uniquely shaped, having a delicately fringed cup framed by five long sepals, and come in shades of white and yellow. This exquisitely different flower may be grown as a pot plant in the North.

ISOLATION WARD.   Place a new pot plant somewhere by itself for a week or so by itself before you put it with your other plants. This will give you a chance to ensure that it is free of insects or disease without running the risk of infecting your other plants.

IVY TREE *(Hedera)*.   Many plants at flowering time change their style of vegetative growth from prostrate to erect. This is common among annuals and perennials, but rare in vines. In ivy, the flowering process triggers the formation of erect branches on which the leaves are different from those on other parts of the vine—narrow and lance-shaped.

Ivy vines flower only when about fifteen years old. The flower head is a cluster of tiny cream colored flowers. If cuttings are made of the erect flowering branches before the flowers are produced, the erect form is retained by the plant that develops from the cuttings.

Take the cutting below the tip at the bud on a young woody stem. Insert cutting into a loose sponge rock (perlite) and peat moss mixture; roots will quickly form. After roots have formed, pinch back the stem to produce an interesting form with side branches. The variegated *Hedera canariensis* with dramatic green and white branches is a wonderful subject for this horticultural wizardry.Remove any leaves that revert to the vine habit.

IXIA *(Ixia hybrida)*.   Ixia, boasting the greatest range of color of any bulb, is cool-loving, which makes it very desirable for indoor cultivation. The stems are slender and graceful; the flowers offer white, yellow, purple, ruby, blue, and green, in many shades and variations, usually with a black eye. The flower spikes contain six to twelve flowers, each one to two inches in diameter. Pot the bulbs in late autumn in a mixture of loam leaf compost and sand, placing eight or nine in a six-inch pot. Keep cool and dark until growth starts, then bring into light and warmth.

J

JACK-O'-LANTERNS ON THE VINE. Pumpkins are fun to grow and fun to eat. In pioneer days they were sliced and hung from cabin roofs to dry for winter storage. They were made into soup, stew, pudding, bread, griddle cakes, and a thick sauce, as well as pie.

Pumpkins are pretty in flower and bright in fruit. To decorate them while still on the vine, start with the seeds. For small pumpkins for small children the variety Small Sugar is best. The variety Jack-O'-Lantern is medium sized; for a truly big one Big Max often weighs in at 100 pounds.

As they ripen they turn from green to yellow orange. While they are still green but have almost reached full size, take a paring knife and carve the jack-o'-lantern face, or a child's name, or any other design, through the rind and into the flesh about ⅛ inch deep.

In a few days a callus will form along the cut lines, and the design will begin to rise up in a distinct pattern, turning light colored against the orange skin, after the pumpkin ripens. No harm comes to the interior of the fruit. Near frost time harvest the decorated pumpkin as usual.

JADE PLANT *(Crassula).* Jade plants have been popular for a long time, and they have been favorites with plant lovers in North America. Some advanced growers are even treating it as a tropical bonsai, yet the jade plant is easy for beginners to grow as well. Common jade, *Crassula argentea*, has dark green leaves that become red-edged in sufficient sun. Mature plants bear clusters of star-shaped white or pale pink flowers at the branch tips in winter or spring. Its dwarf form, *C. argentea minima*, is simply smaller in all its parts.

Jade is virtually trouble-free. Mealybugs are the chief insect pest; eradicate them by using a cotton swab dipped in alcohol. Malathion should not be used on jade or other crassulas.

JAPANESE BEETLE *(Popillia japonica).* This bronze blue, iridescent beetle feeds on all kinds of ornamentals. To control, a rotenone preparation is recommended. In lawns, treat the turf with a bacterial spore dust which will infect the grubs with milky spore disease. Beetles are sometimes caught in traps filled with geranium oil. They may also be lured to feed on trap crops of African marigolds, evening primrose, or woodbine. They are poisoned by leaves of castor bean and blossoms of white geranium.

JAPANESE ROSES. These are the pure rugosa roses that have been developed from species originally found in Japan; sometimes they are known as Japanese roses. They are hardy anywhere and very disease resistant. They should not be confused with hybrid rugosas that have been developed by cross-pollination with other types of roses.

The rugosas are good for a dense hedge and for molding a desired contour in the garden. Their beautiful hips have a higher vitamin C content than any other fruit, even oranges, and the bees hover around the flowers because of their intense perfume. Some kinds include: "Will Alderman," clear lilac pink; "Blanc Double de Coubert," pure white; *Rugosa magnifica*, carmine, and "Frau Dagmar Hastrup" with five petals of clear pink and lower growing than most of this type.

JASMINE (*Jasmine nudiflorum*). The so-called winter jasmine is one of our most brilliant, winter flowering shrubs. The cheerful butter-yellow flowers and red-tinged buds appear in profusion in mid-winter. Before October is out if you grow it on a south or west wall, the earliest flowers will brighten a garden already entering its first bare stages of winter. And then until February or March your winter jasmine should be a never-failing source of blossom for your house. Most generous flowers appear when this hardy plant is grown in a lime-free soil in a position fully exposed to the sun.

JEWELRY. "Beads" from the garden are a delightful bonus from many colorful flowers and their seeds, such as ornamental corn, sunflowers, castor beans, and Job's tears plants. But rose-petal beads with their sweet, mild fragrance, have always been a great favorite. Once these were much in demand for rosaries as well as necklaces and they were very lovely with contrasting mountings of either gold or silver. Here is how they were made.

Rose Beads for a Rosary. In an enamel pan heat 1 cup of salt with 1 heaping cup of rose petals firmly packed. When this has been mashed together, stir in ½ cup of water. Add a drop of oil paint for any desired color or omit if natural color is preferred. Reheat over an asbestos plate, stirring constantly until smooth. Roll out to ¼-inch thickness, cut with a thimble and roll each bead in the palm of the hand until smooth and round. As each bead is rolled, string it on #24 or #26 florist's wire. Hang in a dark place until dry, then string on dental floss. Move beads occasionally while drying to keep them from sticking.

Rose Beads for a Necklace. Put 1¾ cups of flour and 4 tablespoons of salt into a bowl and add a little water to make a smooth dough. Into this press 3 cups of rose petals that have been finely chopped. Flour a bread board and roll the dough to about ¼-inch thickness. Use thimble to cut dough. Roll each circle in the palm of hand to form a smooth bead. Follow above directions for stringing and drying. When stringing add crystal, gold, or silver beads between each rose bead.

A drop or two of rose extract or rose oil will add a delightful fragrance to either rosary or necklace.

JEWELS OF OPAR. This garden treasure has bright, waxy green foliage with myriads of cameo-pink flowers that open every afternoon. It is 1½ feet tall and good to use for borders and rock gardens. It is heat resistant and naturalizes well.

JEWELWEED *(Impatiens biflora* or *I. pallida).* This tender, succulent, tall-growing annual is often found in extensive patches in damp woods. The expressed juice is a light orange color and is an antidote for poison ivy which it often grows near.

The jewelweed has pretty flowers, butter-yellow in color, and these are followed by slipper-shaped seed pods about ¾-inch long. When ripe they will suddenly split, the two sides curling back into tight spirals with an audible snap, and throw their seeds in all directions.

Jewelweed is so called because its leaves are unwettable; rain will stand on the leaves in round drops, shining like jewels, without ever wetting the leaf surface.

JIMSON WEED *(Datura stramonium).* This weed spreads usually by having its seed carried by birds. It is very poisonous, causing a kind of intoxicated state, but has a certain medicinal value. It is helpful when grown with pumpkins.

JOB'S TEARS *(Coix lacryma).* This three-foot ornamental grass bears hard, pearly white seeds which make distinctive necklaces.

JOHNSON GRASS *(Sorghum halepense).* In the West and Southwest this is a particularly troublesome grass. The rhizomes go fairly deep in the ground and are hard to eradicate in garden or flower bed. Geese are helpful, particularly on cotton acreages, and "cotton goosing," as this is called, has expanded rapidly in recent years. White Chinese geese are rented or sold to individual planters for the weeding season.

JOJOBA *(Simmondsia chinensis).* Jojoba is a dense, mounding evergreen desert shrub which may grow as high as eight feet and equally as wide. It is an excellent landscape plant even for a formal garden, for hedges, background, foundations, and screens. Mature plants are hardy to 15°F., but seedlings are sensitive to frost.

JUJUBE *(Rhamnaceae)* CHRIST-THORN *(Zizyphus spina-Christi).* This may be either a shrub or a small tree. They bloom late, the small greenish white flowers giving way to elongated fruits sometimes called "Chinese dates." These can be used in the kitchen or as a tonic food for people and animals.

It is an attractive plant with shiny, bright green leaves and the small, woolly flowers in clusters are richly honey-scented. The fruit is a prized delicacy of the Bedouins. However, the thorns of this shrub dig cruelly into human flesh. Legend says that Christ's crown of thorns was made of its branches. Legend also says that the *Christi* name was given to the shrub because Christ loved its fruits. They are refreshing and alleviate fatigue in the heat of summer.

JULIP MINT. If you have ants in your kitchen, julip mint, a spearmint, or tansy, planted near the kitchen wall or entrance will help keep them away.

JUNGLE CACTI *(Epiphyllums).* These strange flowering plants are almost unbelievably beautiful. Somewhat resembling orchids in their delicate loveliness, they include both day- and night-blooming varieties. They are every color of the rainbow except blue. If you are a cactus lover these will delight you.

To grow these, begin with the right potting mix: 4 parts leaf mold, 1 part fine redwood bark, 1 part well-aged steer manure, 1 part perlite, 1 part horticultural charcoal. (If you cannot obtain leaf mold, commercially packaged camellia-azalea mix is a good substitute.) For each cubic foot of mix, add ½ cup bone meal. If you live in an area which has alkaline water, add ½ cup garden sulphur per cubic foot of mix.

Keep the plant in a pot a bit small for the size of the plant, and do not make the soil too firm. During November and December keep the soil nearly dry (but do not let the skins shrivel); at other times keep it moderately moist. Grow in a temperature of 60° to 70° F. except when resting the plant (when it should be kept at 50° to 55°F.). Give all light possible in winter, but from March through September shade it lightly from strong sun. It may be put outside in summer.

K

KALANCHOE *(Kalanchoe).* These showy, winter blooming plants make long-lived pot plants. Tender succulents with attractive flowers, they are chiefly natives of the Tropics and South Africa and belong to the Crassula family. Sow in spring for winter and spring bloom. The flowers are orange scarlet, pink, orange red, and white.

KALE, FLOWERING. Flowering kale is edible as well as beautiful and very colorful for a special accent plant outdoors. Or it may be grown as a pot plant. Make a flower border of these unusual foliage plants. With cool fall weather the crinkly leaves turn many shades of violet, rose, cream, and green.

KAVA KAVA.   Kava and Ava are the names of two shrubs related to the pepper plant. People have cultivated them for centuries in the Pacific Islands and Australia. The kavas are erect shrubs and may grow as tall as five feet. They have small yellowish cream flowers and round leaves. They may be easily raised in greenhouses and can be grown from stem cuttings.

The roots yield a juice called *kavaic acid*. Peoples of the South Pacific use the roots to make a fermented drink called kava, ava, or kavakava. Kava kava is an herbal substitute which offers many of the same benefits as Yohimbe which is presently restricted by the FDA for use as a nutridisiac. Yohimbe is said to surpass any other substance in its quality and effectiveness as an aid to virility.

The kavas are in the family Piperaceae. The two kinds are *Piper methysticum* and *P. excelsum.*

KERRIA *(Kerria).*   A valuable plant as it flowers well even in dense shade, this small, tough shrub grows upright with thin branches which remain bright green all winter in all but the coldest regions, and even there it grows when given protection. It blooms in mid-May with a wealth of bright 1½- to 1¾-inch bright yellow flowers. And it doesn't stop there—the spring blooming period is followed by light, sporadic flowering through the summer and an impressive show again in early fall.

KIDS.   Teach your children the principles of gardening. They may well be some of the most important skills they will ever learn. To start, keep things simple. Let children grow the veggies and flowers they like. You might even buy them some started plants to ensure the success that builds confidence in growing things. Counsel and encourage, aid and comfort, but don't do their work. And praise their efforts, especially when vegetables or flowers are brought to the table for the rest of the family to enjoy.

KISS-ME-OVER-THE-GARDEN-GATE *(Viola).*   This is the *Viola tricolor* (the wild pansy), sometimes called Johnny-jump-up. It is a hardy perennial, but rather short-lived. Sow the seeds of this charming little flower outdoors; it will need no special care.

KITAIBELIA VITIFOLIA *(Kitabelia).*   This hardy perennial flowering plant with ornamental foliage is a native of eastern Europe and belongs to the hollyhock family, Malvaceae. The stems, which grow about eight feet in height, have large, vinelike leaves; the large pink flowers open in summer.

This plant thrives in a sunny position and is fine for planting at the back of the herbaceous border. It prefers light, well-drained soil, and may be planted in autumn or spring. Propagation is by division of the roots at planting time or by seeds sown outdoors in summer.

KITCHEN. Use flowers to brighten your kitchen. Put them on counter space with overhead lights or under a window. Top off a room divider cabinet with flowers. Light, humidity, and good ventilation are usually in abundant supply, and the proximity of the kitchen sink encourages good watering habits. Keep your green friends well away from cooktops and ovens where heat and fumes may damage them.

Plants enjoy a sudsy bath; wash them with a little mild soap and cool water.

Provided you have a fairly good light where your planter is situated, grow a variety of begonias, spider plants *(Chlorophytum)*, ivies, wandering Jews, ferns, and herbs—and of course grow such vines as grape ivy and philodendron.

Foliage plants will give you more year-round satisfaction than flowering kinds. Try a selection of dieffenbachias, aspidistras, snake plants, ferns, dracaenas, pandanus, and coleus.

Bathrooms provide another opportunity for humidity-loving plants. Choose moisture lovers such as devil's ivy, English ivy, arrowhead vine, hollyfern, and baby's tears. Avoid plants such as cacti that need dry conditions.

KNEE PADS. These can be as simple as iron-on pads, inside or out, for your blue jeans, or a pair of rubber pads with adjustable straps. They make small garden chores like weeding so much easier.

KNIPHOFIA *(Kniphofia)*. This member of the Lily family is sometimes called torch lily or red hot poker. Hailing from Africa and Madagascar, these perennials are also called tritoma. They are very showy plants for borders, mostly in the red and yellow color range. In cold areas they will need winter protection.

KNOTWEED *(Polygonum bistorta)*. The Virginia knotweed can tie a sailor's knot which is put to such a strain when it dries that it snaps, hurling the seeds to germinate as far as possible from the mother plant. At one time the knotweeds had value as secondary feeding plants, particularly in the Asiatic East; others were used for tanning and dyeing (brown, yellow, and indigo). A cousin, the Chinese smartweed, was grown for indigo production. Folklore medicine used the twisted roots of meadow knotweed as a remedy against snake poison. All of this would indicate that the family contains a certain vitality which has never been properly developed except in the case of the buckwheat.

KOCHIA. Burning Bush, Summer Cypress. For a quick-growing ornamental try kochia. It grows thirty inches tall and makes a nice annual hedge. The feathery foliage turns red in the fall.

KOLKWITZIA AMABILIS. Beauty Bush. This is another handsome flowering shrub from China. Its clean foliage is untroubled by insects or diseases. In June the whole plant becomes a fountain of bell-shaped, light pink flowers. It reaches a height of seven to eight feet and will grow anywhere, even thriving in dry, sandy, poor soil.

KUDZU VINE *(Pueraria).* This fast-growing vine is a veritable Jack and the Beanstalk but you must be careful to keep it in bounds or it will completely "take over." The large, splendid foliage is impressive and the plant is excellent for preventing erosion.

L

LABELS. Many flowers when they are not in bloom look so much alike that it is easy to make a mistake when they are dug for replanting or sale. Iris, for instance, are almost impossible to tell apart. Write on the leaves with a magic marker pencil, or for a more permanent label, cut old venetian blinds in one-foot lengths and write the name of the plant with a wax pencil.

When potted, the plant may be inexpensively labeled with a plastic knife. Print out the name of the plant with a name-tape printer, press it onto knife handle, and push the blade into the soil. This is neat, inexpensive, and long-lasting, especially with potted plants.

Plant labels can also be made from 1″ × 3″ strips cut from used bleach jugs. Scratch or write on these with a large needle inserted in a 2-inch section of wood broom handle. Next, rub dark shoe polish over the label, and wipe off excess. This makes the writing both visible and permanent.

Somewhat unusual but neat and practical are the gutter spikes available in 7- or 9-inch lengths from hardware stores. To prevent them from rusting, dip them in rust-resisting enamel and let dry. Then wire on labels with copper wire or plastic twistems and drive the spikes into the ground beside the plants.

LADY'S SLIPPER *(Cypripedium).* These beautiful orchids are found in swamps and wet woodlands, most numerous in the eastern and southeastern states. The flowers that grow on straight stems have a pouchlike lip, hence the name "lady's slipper." They are beautiful, but be careful in handling; all species contain a poisonous substance in the stalks and leaves which frequently causes dermatitis.

LAMB'S-QUARTERS (*Chenopodium album*) PIGWEED.   Lamb's-quarters is a close relative of garden spinach. Though considered a weed it is good to eat and is actually richer in vitamin C and far richer in vitamin A than spinach. While not quite as rich as spinach in iron and potassium, it is still a good source of these important minerals. But the area where it excels is as a source of calcium. Those 309 milligrams of calcium per 100 grams make this the richest source of calcium found among green, leafy vegetables; green amaranth is second with turnip greens third.

Lamb's-quarters gives added vigor to zinnias, marigolds, peonies, and pansies when grown nearby. It is also a soil improver. But the weed does harbor the leaf miner – this is offset by the fact that it plays host as well to the beneficial lady beetle.

LAVENDER (*Lavendula*).   Lavender has a long and creditable history as a stimulating or medicinal plant. Among its many virtues is its soothing effect on the stomach, its use as a disinfectant, and its power to relieve sprains, headaches, and toothaches. Flies steer clear of it and, according to the old demonologists, the fragrant odor of lavender is guaranteed to ward off evil spirits. The essential oil of lavender is produced from the leaves and has been an ingredient of love philtres from earliest times. The oil also stimulates the generation of new cells and in so doing helps to preserve the health and youth of the skin.

Three varieties used in industry, medicine, and household preparations, are beautiful additions to the garden. These are spike (*L. spica*), true or English (*L. vera*), or French lavender (*L. stoechas*). Both flowers and leaves are fragrant. Plants must be grown in poor soil to produce the most fragrance; in good soil they grow more luxuriantly but fragrant essential oils are lacking.

LAWN.   Lawns are the greenest show on earth. Where the flower beds are the picture, lawns are often the frame. A lawn is a great asset to cooler living, for an acre of grass in front of your home gives off 2,400 gallons of water every hot summer day. This has a cooling effect equal to a 140,000-pound air conditioner which amounts to a seventy-ton machine. Along with trees, shrubs, and flowers, a lawn helps purify the air, putting oxygen back into it so we can breathe much easier.

If you live in a dry location and have difficulty getting grasses to flourish, consider establishing a lawn of fragrant camomile (*Anthemis nobilis*). Drought has little effect on this green plant. Sow it just like grass. When cut it will give out a fragrant odor. To get it started in spring, sow a mixture of camomile and lawn grasses together; as the season advances the strong growing camomile will take over the lawn as the grasses begin to lose ground.

LAWN GRASSES. Here is what a good lawn grass needs to grow:

1. A fertile topsoil, deep and rich in humus, at least six inches.
2. Cutting at a reasonable height. During the hot summer months, when weeds are a problem, trim your grass as high as three or four inches. This is particularly helpful to perennial ryegrass. At other times of the year grass should be two inches tall after mowing; it's best equipped to resist disease at this height.
3. An occasional thorough watering.

Crabgrass, one of the worst pests, and good lawn grasses thrive on a different set of growing conditions. Encourage or discourage the weed by altering its environment.

1. Crabgrass doesn't grow as strongly on good soil as do good lawn grasses. It cannot compete with ryegrass and bluegrass when those plants have a good soil to grow in.
2. Crabgrass likes to grow close to the ground; therefore it is not stunted by low mowing. On the contrary, low mowing gives it the room it needs to spread.

LEAFHOPPER. The leafhopper is repelled by petunia and geranium.

LEAF ODORS. Leaves hold their aromatic scent far longer than flowers. Often they are sweeter in a dried state than when fresh. To release the odor of herbs and leaves, grind them in a mortar using a pestle. A heavy and very beautiful mortar and pestle made of marble can be found at Lillian Vernon's (See Sources of Supply).

LICORICE (*Glycyrrhiza glabra*). The wild licorice of North America is *G. lepidota*. From licorice roots we get a valuable flavoring material which is fifty times as sweet as sugar. But oddly, although sweeter than sugar, licorice has the power to quench thirst. In 1951 it was discovered that licorice root contained the female sex hormone estrogen, used in the treatment of menopausal problems.

Chewing a licorice stick helps those who wish to stop smoking. Even the fibers which remain after the licorice is extracted from the roots are valuable. They are used in making fire-fighting foam, boxboard, insulation board, and other products.

LIGULARIA DENTATA. This bold perennial has large, heart-shaped leaves, bronzy green on the upper surface and a rich mahogany red below. In July and August, orange daisy flowers brighten the plant. It does best in a constantly moist soil in a partially shaded location.

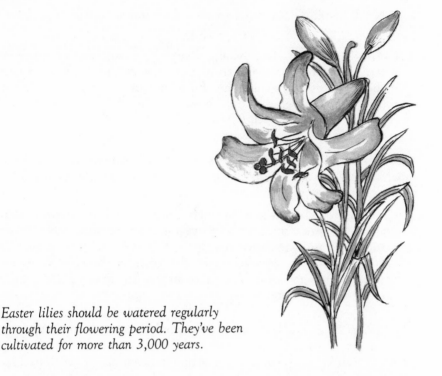

Easter lilies should be watered regularly through their flowering period. They've been cultivated for more than 3,000 years.

LILY *(Lilium).*   This perennial comes in many colors and is lovely used with delphinium, asters, and marigolds. It grows well in raised beds and is ideal for mass plantings along a wall. *L. longiflorum,* the Easter lily, is pure white with a powerful fragrance, and is a very popular gift at Easter time.

LILY OF THE VALLEY *(Convallaria majalis).*   Lily of the valley, a single species, is a lovely, fragrant, spring-blooming herb. It is commonly cultivated in the partial shade of gardens everywhere and is a great favorite. It looks best in mass plantings but makes a fine ground cover for narcissus. However, if narcissus and lily of the valley flowers are put together in the same vase, they will soon wither.

To get flowers by Christmas, professional Dutch growers of the 1870's planted moss-wrapped pips (a pip is a rooted bud arising from the rootstalk) in sand about a month before Christmas, giving them bottom heat and liberal watering until sprouts appeared.

The leaves, flower, berries, and rootstalk are well known for their toxicity. They contain dangerous amounts of cardiac glycosides (convallarin and convallamarin). Children have lost their lives just by drinking the water from a vase containing a bouquet of lily of the valley.

LINNAEUS, CAROLUS *(Karl von Linnle).* Linnaeus, the great Swedish naturalist, was born in 1707. He composed a floral clock to determine the time of day by the opening and closing of certain flowers which he observed folded and unfolded their petals at regular hours. A few of these which served for the construction of his dial were:

Dandelion opens from 5 to 6 a.m. and closes between 8 and 9 p.m., mouse-ear hawkweed opens at 8 a.m. and closes at 2 p.m., yellow goat's beard opens at sunrise and shuts at noon, smooth sow thistle opens at 5 a.m. and closes between 11 a.m. and noon, cultivated lettuce opens at 7 a.m. and closes at 10 a.m., white water lily opens between 5 and 6 a.m., mallow opens at 9 to 10 a.m. and closes at 1 p.m.

LIVE FOREVER *(Sedum spectabile).* This hardy succulent lives up to its name. It displays heads of pinkish bloom above gray green foliage in late summer. The plant will tolerate poor growing conditions, is transplantable at almost any season, and propagates readily by leaf cuttings. Recent varieties have flower colors ranging from ivory to rosy red.

LOBELIA *(Lobelia inflata)* BLUE CARDINAL FLOWER. This field plant of North America is commonly called Indian tobacco. The leaves are pointed and yellowish green with hooded flowers of a brilliant blue. This is one of the most important herbs of the American Indian, belonging to that small group which are virtually cure-alls; it is beneficial to the whole body. It is useful as well as an external application for many types of sores.

Since the lobelia blossom is such a lovely blue, it combines well with white alyssum and red rose buds for bouquets or tiny place markers. And lobelia, like alyssum, is another excellent plant for "clothing" a pot, planter, or window box planted with larger plants. Keep well pinched to promote shapeliness and persistent bloom.

LOOK ALIKES. Virginia creeper is frequently mistaken for poison ivy, which it does somewhat resemble. Yet its five leaflets, compared to only three for all forms of poison ivy and poison oak, make it easily recognizable. Other innocent plants that sometimes suffer from an identity crisis are the harmless Boston ivy *(Parthenocissus tricuspidata)* and marine ivy *(Cissus incisa).*

The real culprits to watch for are:

Poison Ivy *(Rhus radicans).* Usually this is a vine, but may, sometimes, especially in open sun, be a shrub from a few inches to several feet high.

Poison Oak *(Rhus toxicodendron).* Found in the South, it is sometimes called oak-leaf poison ivy. It is shrubby with leaflets covered by downy hairs and lobes resembling small oak leaves.

Be careful around these two, for even particles of the irritant oil wafted through the air on smoke or pollen can affect the eyes or lungs of allergic people. Jewelweed, which often grows nearby, is very good to use as a remedy against poison ivy. It relieves the itching almost at once. Boil a pot of the jewelweed and strain the juice. Keep juice refrigerated or freeze cubes of it and bag them for future use.

LOVE-IN-A-MIST *(Nigella)*.   Whether you think of it as love-in-a-mist, or devil-in-a-bush, its other name, this charming blue, purple blue, or white annual makes a hardy border plant where it is very beautiful for several weeks. Well-grown plants bear finer blooms and remain in flower longer than those which are crowded.

LOVE-LIES-BLEEDING *(Amaranthus caudatus)* TASSEL FLOWER.   This native of India, the Philippines, and other warm countries has drooping stems bearing dark, reddish purple blooms.

LOW LIGHT AREAS.   A dim location is the perfect spot for a dieffenbachia, a tropical plant from South and Central America. It can survive on only two hours of sunlight a day but must have humidity. The plant grows well grouped with palms, aglaonemas, ferns, and dracaenas.

LUFFA GOURD *(Lageneria)*.   Luffas need hot temperatures and a long growing season. The fibrous, spongelike interior of the fruit is used for bathing.

LUPINE *(Lupinus)*.   There are about 100 species of lupines found in the United States. They are showy, hardy, and grow profusely in fields, on ranges and mountainsides. The pealike blossoms in loose clusters at the ends of the branches, may be blue, purple, yellow, pink, or white. Some but not all species of lupine, hold toxic alkaloids throughout the entire plant and are a common cause of stock poisoning.

Lupine, which is a legume, is helpful to the growth of corn as well as most cultivated crops. Plant in full sun or light shade.

Cultivated lupines come in a gorgeous assortment of colors, blues, pinks, reds, purples, maroons, and many striking bicolor combinations. The individual flowers are large with some standards up to an inch across.

M

MAGNOLIA *(M. stellata, M. soulangeana)*.   Magnolias are of special interest because they have the largest flower of any tree in our gardens. The lustrous evergreen leaves, the big deliciously fragrant white blossoms, the conelike fruits that flush from pale green to rose, all have helped to give the magnolias a preeminent place in every country where ornamental planting is valued.

Magnolias are reasonably hardy and in sheltered locations may be planted as far north as Massachusetts. They prefer a rich, moist soil. Transplanting is, however, a difficult operation and is best done when new growth starts. The flowers show up marvellously against a dark background of evergreens.

MAHONIA (*Mahonia*).   These evergreen shrubs have compound, hollylike leaves, fragrant yellow flowers, and berries that are both edible and delicious. Plant mahonia for foundations, background, screens, and ground cover.

MARIGOLDS (*Tagetes*).   Marigolds are as helpful in the flower garden as in the vegetable garden, serving the same purpose of driving away nematodes. They are particularly useful with chrysanthemums, calendula, and dahlia. Brown areas on lower leaves signal the underground feeding of the foliar nematodes.

Snowbird marigold, American carnation flowered (Burpee) is the pristine white marigold and the first white variety to be developed from the seeds. It is a shimmering beauty with double flowers 2½ to 3 inches across. Of uniform habit, Snowbird grows about 18 inches tall.

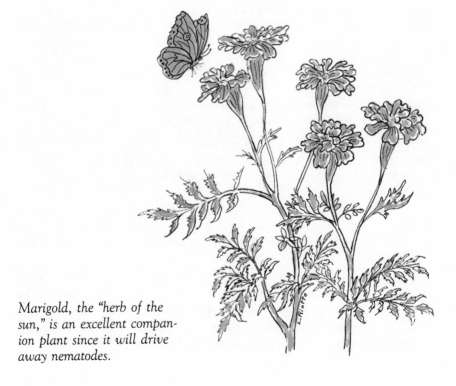

Marigold, the "herb of the sun," is an excellent companion plant since it will drive away nematodes.

MARTYNIA *(Martynia)* DEVIL'S CLAW, UNICORN PLANT. Sonoran tribes domesticated this plant for the long black fibers of the "claws" on the fruit. They were used in basket weaving. The seeds may be eaten like sunflower seeds or pressed for oil. The fuzzy green seedpods are picked when half-grown if wanted for pickling. The strange mature pods are often used for decorations.

MEADOWSWEET *(Filipendula ulmaria, Spiraea).* Meadowsweet is perhaps the most important of all nature's remedies. Chemists discovered from meadowsweet, acetyl salicylic, synthesized it, and called it aspirin. But the dried leaf in its natural form can be ground with mortar and pestle and used wherever aspirin is recommended without aspirin's side effects. Meadowsweet is particularly recommended by herbalists as an antidote for rheumatism, arthritis, gout, and all kidney and bladder complaints, particularly gravel and cystitis.

MEALYBUGS. Live controls for mealybugs include cryptolaemus beetles and lacewing larvae. On houseplants a small watercolor brush dipped in alcohol will kill them. More pliable than Q-tips, the brush will not bruise the plant.

MESCAL BEAN *(Sophora secundiflora).* The beautiful green, glossy leaves, and floral display of fragrant sweet-pea-like lilac flowers, borne in a drooping inflorescence, are very impressive. This excellent landscape plant is particularly good for poor alkaline soils, and tolerates heat and drought well.

The beautiful seeds are large, brilliant coral and, even though they are very poisonous, the Mexicans use them for necklaces. Powdered seeds of this flowering shrub are toxic to armyworms.

MESCALINE or PEYOTE. This drug is obtained from a small cactus plant that grows in the Rio Grande region of Mexico and the United States. Indian tribes of the region use it for medicinal and religious purposes. Mescaline produces visions, often with brilliant color. It also produces psychological disturbances and trances. Psychological investigation indicates that mescaline causes confusion of personality and a sense of unreality similar to some kinds of mental illness. No use has been found for mescaline in modern medicine, and it is considered to be a dangerous drug.

MESQUITE *(Prosopis)* HONEY POD. These interesting and beautiful multiple-trunked deciduous trees or large shrubs are found growing in desert areas. The clusters of creamy white flowers attract bees which make an excellent honey from their nectar. The fine-textured, fernlike foliage gives light shade. These plants grow slowly in nature but faster in cultivation. They do well in lawns and are nitrogen-fixing members of the bean family.

The mesquite is almost an object of worship to desert dwellers. The long, fat pods supply Mexicans and Indians with a nutritious food. Cattle thrive on the young shoots when other forage is lacking. The deep-reaching roots, sixty feet or more in length, are hauled out of the ground for fuel, posts, railroad ties, furniture, and paving blocks. The wood is also cut into building and fencing materials, two great needs of the desert.

MEXICAN BEAN BEETLE.   The beetle is repelled by marigold, potato, rosemary, summer savory, and petunia.

MEXICAN HAT *(Ratibida columnifera forma pulcherrima)*.   This dark mahogany red, or yellow with a red blotch, flower is a form of the prairie coneflower, the dramatic yellow daisies with cone-shaped brown centers. The perennial plant is about one foot tall, and given full sun it will bloom continuously from summer to fall. Grow it with the taller (three to five feet), purple coneflower *(Echinacea purpurea)*, which also flowers from summer to fall. The lavender purple daisies make excellent cut flowers.

MIGNONETTE *(Reseda odorata)* Little Darling.   The mignonette has a low bushy mass of smooth, soft green leaves. The tiny flowers growing on tall spikes, are yellowish white with reddish pollen stalks inside. The larger flowered varieties are prettier but less fragrant. The mignonette is a good border plant which grows best in cool temperatures and light soil. If the seed fails to sprout, check for ants; they have been known to carry off the seed.

MILK.   It is sometimes helpful to spray milk over plants suffering from mildew or mold, or tobacco mosaic.

MILKWEED *(Asclepias)*.   These rather coarse erect plants grow in dry fields, on hillsides, in woods, and along roadsides. Their profuse milky juice accounts for their name. Milkweed juice is said to remove warts.

There are about sixty species distributed throughout the United States but only a few are cultivated. The white, pink or rose colored flowers develop in round clusters. The large, rough-surfaced, flat seed pods are filled with many seeds, each with a tuft of long, silky hairs. Divested of the seeds, the silky hairs are sometimes used as picture background for butterflies and dried flowers.

A method of trapping cutworms used at one time consisted of placing compact handfuls of milkweed in every fifth row or hill of cultivation and tamping them down. Cutworms gathered in this trap material where they could be easily collected. Clover and mullein were also used for this purpose.

Some species of milkweed have medicinal value, but all are known for their content of resins and most of them are exceedingly poisonous to humans and livestock. The poison is concentrated in the stout stem and leaves.

MILKY SPORE DISEASE. Milky spore disease to combat Japanese beetles has been isolated and is now manufactured and distributed commercially. It is selective and the disease affects the insect in the grub stage; it will not affect any species other than the Japanese beetle.

MIMOSA (*Mimosa pudica*) HUMBLE PLANT, TOUCH-ME-NOT, HAIRY SENSITIVE PLANT.   The sensitive plant is so called because its leaves will fold together with sufficient irritation or cloudy weather, representing one of the most remarkable cases of physiological response in the plant kingdom. The timorous plant has a mechanism which reacts whenever a beetle, ant, or worm crawls up its stem toward its delicate leaves; as the intruder touches a spur, the stem raises, the leaves fold up, and the assailant is either rolled off the branch by the unexpected movement or is obliged to draw back in fright.

MINERAL DEFICIENCIES.   Elements considered secondary such as calcium, boron, silicon, and manganese have been shown to exert appreciable influence on plant diseases. Home remedies of experienced gardeners may be countering the mineral deficiencies of plants. These include rusty nails around the roots of roses to keep the plants healthy and free of insects and mildew. The addition of rusty nails, iron filings, and other rusty material to the soil has been known to change the blossoms of wisteria from white to purple.

MINI-GREENHOUSE.   To root African violets, rex begonias, roses, and small evergreen cuttings, use a clean, label-free, three-pound peanut butter jar turned upside down on its lid for the greenhouse. Fill the lid with moist gravel. Then place a pot containing the moist, sterile, rooting medium and cuttings on the gravel. Tip out any excess water. Twist jar down over the lid to seal.

MINTS (*Mentha*).   There are many delightful mints used medicinally, for cooking, and for fragrance. These seven are among the most frequently grown:

Orange mint or **bergamot mint** (*M. citrata*) has lavender blossoms in dense flower spikes and a characteristic minty odor.

Golden apple mint (*M. gentilis*) has smooth, deep green leaves, variegated with yellow. It grows about 2 feet and makes an attractive ground cover where taller spring-flowering bulbs are planted.

The leaves of peppermint make a wonderful tea, reputed to be good for the nerves. One author has called this tea one of nature's best tranquilizers.

Peppermint (M. *piperita*) or its flavor is familiar to many people. It has strongly scented small purple flowers and 3-inch leaves with toothed edges. Peppermint grows to about 3 feet.

Pennyroyal (M. *pulegium*) is an attractive plant with small, rosy lilac flowers, blooming late in the summer and early autumn. Pennyroyal is believed to repel insects in the garden and is good to rub on a cat's collar to repel fleas.

Jewel mint of Corsica (M. *requienii*) is a creeping sort that grows only about 1 inch high. The tiny, round leaves form a mosslike mat. Small, light purple flowers appear in summer. When bruised or crushed underfoot the foliage has a delightful minty or sagelike fragrance. Plant it between stepping stones.

Apple mint (M. *rotundifolia*) has stiff stems growing 20 to 30 inches tall. The rounded leaves are slightly hairy and gray green, about ¼ inch long. This mint has purplish white flowers and is not good for culinary use, but American apple mint (M. *gentilis* var. *variegata*) has a fruity, refreshing odor and taste.

Spearmint (M. *spicata*) is another familiar species. It is usually used in mint jelly. The leaves are dark green; the mature plant is 1½ to 2 feet tall.

Mints have been used medicinally since ancient times and modern commerce still makes use of them. Spearmint and peppermint are two of the most common flavorings for everything from chewing gum to toothpaste.

Most mints do best in light, moderately rich soil that is moist, and in shade or partial shade. They spread underground by means of stems and runners. Use the leaves fresh or dried; add them to potpourris, lamb, and jelly; spearmint is the best for garnishing iced drinks (the mint juleps of the South); add fresh leaves of peppermint, pineapple, apple, and orange mints to fruit cocktails or sprinkle over ice cream.

MITES. Live controls of mites include lacewing larvae and mite-eating mites *(Phytoseiulus persimilis)* which may be purchased from commercial insectaries. You may also try dusting sulfur (if temperature is under 90°F.), being sure to cover undersides of leaves. Mites are repelled also by onion, garlic, and chives.

MONEY PLANT *(Lunaria).* This purple or white biennial is best as a filler plant until fall when the silver pods begin to show at their best. Use dried in a winter bouquet or mixed with other dried material.

MONKEY FLOWER *(Mimulus).* These perennials with large, showy, snapdragonlike flowers, like to grow beside streams and ponds, even swampy land. The bush monkey flower *(Mimulus diplacus)* is a flowering shrub of the snapdragon family and grows to 4 or 5 feet. It blooms over a long period with flowers in shades of orange, yellow, and red. Grow in full sun in well-drained soil; the plants are drought tolerant.

MOON DUST. Some of the most exciting seed sowing in history is that done in recent years in moon dust brought from the Apollo missions. Cabbage, Brussels sprouts, broccoli, carrots, lettuce, and radishes were sown in material in which moon dust had been incorporated. These were compared with others similarly grown in earth dust. For reasons as yet unclear, the plants grew larger in the moon dust. It was shown also that the plants extracted moon nutrients that were insoluble in water, without any help from micro-organisms in the soil. Until now, we have supposed that soil fungi and bacteria had a vital role in unlocking the soil minerals plants use, but this experiment reopens that matter, since apparently the plants alone somehow dissolved the minerals for their use from the lifeless lunar dust.

MOONFLOWER *(Ipomoea pandurata)* MAN-OF-THE-EARTH, WILD POTATO. This is a very hardy tuberous vine with flowers similar to morning glory but larger. They have a delicious fragrance, slightly reminiscent of lemon, fresh and clean.

MORMON CRICKET SPORE *(Anabrus simplex Halderman).*   Mormon Cricket Control uses a natural spore that disables the Mormon cricket without affecting other insects, animals, or man. And it will not affect the California sea gulls which feed upon the Mormon cricket. It is a safe biological control alternative that will not pollute the environment, and may be used indoors or out for treating lawns, gardens, and basements. Also the spore will not harm infants or pets. (Mellingers)

MORNING GLORY *(Ipomoea purpurea).*   This popular favorite possesses a simple beauty, especially in its modern versions of Heavenly Blue, Pearly Gates, and Scarlet O'Hara (Burpee). Rapid in growth it must be provided with something to twine about; if not it will twine on whatever is nearest, no matter what it is. Morning glory is a great success in a window box. With light support it should reach the ceiling by midsummer, blooming every foot of the way. It is great for covering trellises and arbors and for hiding unsightly areas. Morning glory is said to stimulate the growth of melon seeds.

MOSQUITOES and **MIDGES**.   To keep these away when spending evenings outdoors, gather aromatic herbs such as sage, rosemary, elcampane, and others; add some dry grass or paper; place in a rather large open can (or cans); sprinkle the herbs with paraffin; and ignite. The pungent and pleasant smoke will clear the air of mosquitoes and other biting insects. Mexican Indians burn thuja pine in this way.

MOTHER-OF-THOUSANDS *(Saxifraga sarmentosa)* Strawberry Geranium, Aaron's Beard.   Many admirers have grown mother-of-thousands in hanging baskets or window boxes. Prettily colored, the leaves are light green, variegated with silver above and reddish on the undersides. The flower stalk rises about a foot high and produces white flowers in loose panicles.

From the rosette of leaves come runners which, as they touch moist soil, root and produce new plants. When a young plant acquires six leaves, it may be broken off from the parent and started on its own. This plant grows best in rich, sandy soil with a little filtered sunlight. Mother-of-thousands may be planted outdoors and will survive mild winters, even in the vicinity of New York.

MOTH REPELLENT.   If you detest the odor of commercial moth repellents, try this one from Euell Gibbons' *Stalking the Healthful Herbs.* One pound pine needles (the needles of western pinon pine are best, he advises), 1 ounce cedar shavings, and ½ ounce of shavings from the root of sassafras.

Line a drawer with paper, sprinkle in the mixture, and cover with a thickness of cloth, something like a piece of an old sheet or a thin bath towel, fastening it firmly in place with thumb tacks. Store your woollens on top and they will not only be protected from moths, but have a clean, fresh fragrance when you take them out to wear. Feverfew, sage, tansy, and members of the Artemisia family contain camphor and are also moth repellents.

MULCH. A good mulch can often double the time a flowerbed can go between waterings. Organic mulches act as insulators because of their low heat conducting properties. A mulch such as pea gravel is a good heat conductor. Therefore dark gravel mulches tend to warm light colored soils and organic mulch tends to keep soil cool. Use this principle to slow down or speed up plant growth in the spring. Organic mulches also decrease weed seed germination. Mulches are attractive. They also reduce splashing of mud onto flowers and foliage, again reducing possible disease development.

MULLEIN (Verbascum thapaus) FLANNEL LEAF, BEGGAR'S BLANKET, ADAM'S FLANNEL, VELVET PLANT, FELTWORT, BULLOCK'S LUNGWORT, CLOWN'S LUNGWORT, CUDDY'S LUNGS, TINDER PLANT, RAG PAPER, CANDLEWICK PLANT, WITCH'S CANDLE, HAG'S TAPER, TORCHES, AARON'S ROD, JACOB'S STAFF, SHEPHERD'S CLUB, QUAKER ROUGE. For sheer diversity of names, this one wins the prize.

Mullein is valuable in alleviating human ills and has long been known to many people of varying cultures, all of whom agree on its value as a healing herb. The fresh and dried leaves and the fresh flowers are the parts used in home remedies. Officially, mullein has been recognized as a valuable demulcent, emollient, expectorant, and mild astringent.

N

NARCISSUS (Amaryllidaceae) DAFFODILS, JONQUILS, PAPER-WHITES. This lovely perennial in solid borders and masses is one of the first flowers to welcome spring. The flowers may be yellow, yellow with white, orange, pink, apricot, white, or cream. But beware, the bulbs are poisonous and must never be eaten.

Sow African marigolds (Tagetes erecta) before planting narcissus bulbs to defeat certain nematodes which usually attack the bulbs. Sulphur-containing substances called *thiophenes* are present in root exudates of African marigolds, as well as in many other plants from the Compositae and Umbrelliferae families; these repel the nematodes.

NASTURTIUM (Tropaeolum majus). Forms of this South American tendril climber brighten a fence or trellis with brilliant shades of yellow, orange, and red. It flowers best in full sun and lean soil. Potash added to the soil aids bloom development. Nasturtium, an old Latin word used by Pliny, was derived by him from *narsus*, the nose, and *tortus*, twisted, in reference to the supposed contortions of the nose caused by the hot, pungent odor and taste of these flowers. Nasturtium and rose geranium in a nosegay complement each other.

Nasturtium is an amiable flower which gets along in the worst sort of soil. It was an inhabitant of many an early American garden, being ideal for bouquets, and the flowers, leaves, and seeds were eaten. Flowers and leaves went into salads, and the seeds were pickled after being picked green; so treated, they were thought to be a nice substitute for capers.

Sow nasturtium seed around apple trees in spring to combat the woolly aphis. Sow a few in each hill to repel cucumber beetles. Sow with broccoli against aphis. Though nasturtium often have aphis of their own, they seem to keep them away from their companion plants.

NEMATODES. Nematodes are repelled by marigolds (both African and French), salvia (scarlet sage), dahlia, calendula (pot marigold), and asparagus.

NETTLE *(Urtica diocia)* STINGING NETTLE. This is considered one of the dynamic plants and is especially good when used in the compost pile. Its name *diocia* indicates that the staminate and pistillate flowers, instead of being together on the same plant as usually found, are on separate plants. Therefore look for seed only on some, not all, nettle plants. Both flowers and seed are greenish and inconspicuous. The stinging hairs cover practically the whole plant and prick the skin with their tiny silica tips, letting in enough formic acid to be felt. Even so, the young leaves are very good to eat, though you must wear gloves when picking them. Wash and cook quickly; no other plant has as much iron as this one.

Nettles help plants withstand lice, slugs, and snails in wet weather. They also strengthen the growth of mint and tomatoes and increase the aromatic qualities of many herbs such as sage, peppermint, marjoram, angelica, and valerian. Fruit packed in nettle hay will be free of mold and keep longer.

NIGHT-BLOOMING CEREUS *(Cereus grandiflorus)*. No mention of night-blooming flowers would be complete without this one, a strange cactus of the West Indies. Its bristling tortured stems give birth in the darkness to the most spectacular of blossoms.

Truly a night bloomer, often at midnight, this lovely flower with the saxaphone stem also emits a delightful fragrance. The waxy white blossoms with delicate pink sepals are breathtakingly beautiful. Once it has bloomed, however, the flower's sense of humor seems to assert itself, the spent blossoms hang limply, looking a bit like the legs of a freshly plucked chicken, the rose sepals descending like bedraggled feathers from the yellow tubular stems.

NIGHT-BLOOMING JASMINE (*Cestrum nocturnum*).   This member of the nightshade family is cultivated for its fragrant night-blooming, trumpet-shaped flowers. The leaves are somewhat oval; the fruit, following the flowers, is small and berrylike. All cestrums are extremely poisonous if eaten.

NITROGEN-FIXING PLANTS.   Colonies of nitrogen-fixing bacteria form on the roots of members of the bean family. These organisms take nitrogen gas from the air and convert it into nitrates, which remain in the soil and enhance it. It is far better to encourage these minute creatures to manufacture nitrates than to apply nitrogen-containing chemical manures. They charge nothing for their work and their own death makes the earth more productive. To increase nitrates in the soil, grow peas and beans in different parts of the garden each year.

Since the entire family of Leguminosae has the same properties, grow sweet peas and lupins in the flower garden for the same reason. Never, when clearing ground in autumn, burn the roots of leguminous plants. Break them up instead and add them to the compost heap. The tiny nodules on the roots contain the nitrogen.

O

ODOR. See FRAGRANCE.

OFFICE PLANTS.   Increasingly business offices are making use of green plants. Professional space planners and interior designers are entering the picture. Their first concern is function and efficiency, but they are making very effective use of healthy, thriving plants to soften the lines of functional architecture and at the same time, subtly direct traffic, diffuse sound, or screen certain areas.

Selecting the wrong plants can be an expensive mistake, so consulting an interior landscape specialist is often money well spent. It may be best to not only let the specialist provide the plants but maintain them as well.

For the small office the tried and true sanseveria, rubber plant, cactus, African violets, and ivies are good choices.

OFFICINALIS.   Any plant with "officinalis" as part of its name is listed in the official pharmacopoeia, a book containing standard formulas and methods for the preparation of medicine, drugs, and other remedial substances. The plants listed may vary from country to country. For instance, the British pharmacopoeia lists some plants which are not found in the American. And some plants formerly listed in the American have been deleted down through the years.

OKRA *(Hibiscus esculentus)*. This is one of our most beautiful and stately garden vegetables with large, soft yellow blossoms beloved of bumblebees. Okra is the basis for gumbo, famous as a love food since African slaves developed it when they were brought to Louisiana. Blacks brought the plant to New Orleans and there concocted their gumbo stews, using its young pods and gumbo file, a thickening agent made from sassafras leaves. Crabs, oysters, ham, chicken, and many other tempting foods are combined in gumbos. Southern-fried okra dipped in cornmeal is another true "soul food."

OLDER CITIZENS WITH GREEN THUMBS. Some retired persons are born gardeners, some learn gardening after they retire, still others have gardening thrust upon them. Many retirees have moved to places where they must learn about new soils, new climates, and even new types of plants. For born gardeners the new environments present an interesting challenge.

Those who have never gardened before become gardeners as the result of encouragement by neighbors, friends, and relatives. And then there are those who have had gardening thrust upon them as a way to pass the time, or because social pressure has demanded that they "keep the place looking nice." If you asked these seniors why, they would probably say because they like growing things, because it takes them outdoors and gives them exercise, or because it's something pleasant to do.

Even the handicapped find a way.

OLEANDER *(Nerium oleander)*. Beautiful but poisonous, this houseplant may be grown outdoors in the South. It makes a shrub about fifteen feet tall, with leathery lance-shaped leaves and showy, roselike flowers in red or white. Oleander is easily grown from cuttings. All parts of the plant are poisonous but effective against codling moths.

Japaca, yellow oleander *(Theretia peruviana)*. All parts except leaves and fruit pulp are used to make a cold water extraction effective against a number of insect pests, especially aphids.

OPHIOPOGON PLANISCAPUS. This is a very unusual plant because of its black leaves. It makes a splendid ground cover, spreading slowly to form a dense turf. The purple black leaves accentuate the bright colors of neighboring plants. It is very effective with yellow-leaved companion plants which glow like a beacon in contrast. Its own small pink flowers are followed by glossy black berries.

ORCHID CACTI *(Epiphyllums)*. These incredibly beautiful plants are the so-called jungle cacti or leaf-flowering cacti. The name Epiphyllum is derived from *epi*, upon, and *phyllon*, a leaf, and refers to the location of the flowers. Epiphyllums may be grown in baskets suspended from the greenhouse roof. They require:

Light. Filtered light preferred, *never* direct noonday sun.

Humidity. Approximately 50 percent. Mist during summer months.

Temperature. 45 to 70 degrees F. preferred greenhouse temperature. Protect from frost.

Watering. Water when soil surface has dried to a depth of 1½ inches. Water less frequently during winter.

Potting Mix. Must be coarse and fast draining. Packaged indoor planter mix if coarse enough may be used. Or a mix of the following proportions: four parts leaf mold, one part perlite, one part medium bark, one part horticultural charcoal.

Fertilizer. Mild fertilizer (with no higher than 10 percent nitrogen), once a month starting in April, ending in fall. Once in February and again in November apply low nitrogen fertilizer to promote blooms.

Blooms. Normally occur on two- to three-year-old, root-bound plants.

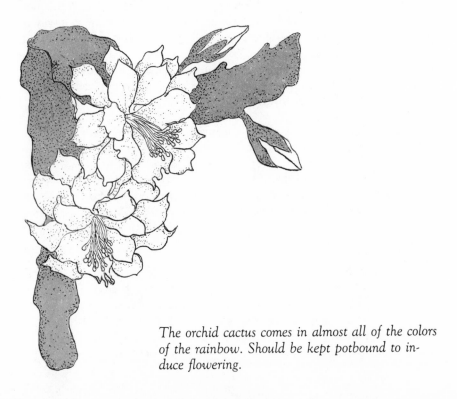

The orchid cactus comes in almost all of the colors of the rainbow. Should be kept potbound to induce flowering.

This pink lady's slipper is a North American woodland orchid, and is as lovely as its many tropical relatives.

ORCHIDS *(Orchidaceae).*   All members of the family look a bit like the gorgeous flowers found in greenhouses and florists' shops. The family relations number more than 6,000 species and many are dainty wild flowers that grow in cool, damp American woods and swamps. These include the white, yellow, and purple lady's slippers; the calopogon; the violet pink arethusa; the calypso; and the fragrant, pale pink moccasin flower.

In the tropical countries many orchids are air plants, attaching themselves to the bark of trees and sending roots into the air from which they take their food.

The orchid takes many unusual forms; the blossoms of one species look like butterflies. Another species furnishes vanilla. The tubers of still another are dried for their nourishing starch. They are sold on the market as "salep," which is used in medicine as a lubricant.

Orchids have the longest blooming period of any flowers. The blossoms of certain kinds may remain open for five weeks or more.

The largest orchid is *Grammatophyllum speciosium*, native to Malaysia. A specimen recorded in Penang, West Malaysia in the 19th century had thirty spikes, the plant was eight feet tall with a diameter of more than forty feet. The largest orchid flower is that of *Selenipedium caudatum*, found in tropical areas of America. Its petals are up to eighteen inches long with a maximum outstretched diameter of three feet.

ORGANIC MATTER. Besides its gradual release of nutrients needed for plant growth, organic matter is valuable for several other reasons:

1. It improves soil aeration.
2. It improves the water-holding capacity of the soil.
3. It reduces soil crusting.
4. It stimulates the growth of beneficial microorganisms, some of which may destroy harmful microorganisms or prevent their growth.
5. It assists nematode control by supporting parasites, predators, and diseases of nematodes.

OSAGE ORANGE *(Maclura pomifera)* BODARK, BOIS D'ARC, BOXWOOD. The name refers to the Osage Indians who used its wood for bow-making. The plant grows wild in Oklahoma, Texas, and Arkansas. The yellow fruit looks much like an orange but is inedible. Cut and oven-dried, the fruit is sliced and used to make lovely flowers (which may be painted) for decorations. The pioneers planted Osage orange for a living fence around their farms before barbed wire came into general use. Posts sprout easily and soon become trees. The wood is also used for making wagon wheels. A yellow dye is made by boiling chips in water.

Roots, wood, and bark repel insects, particularly crickets and roaches.

OSMOSIS. This is the passage of one fluid into another through a membrane between them. It occurs with both liquids and gases. This passage, or transfusion, results in a mixture of the two fluids. Osmosis takes place through a semipermeable membrane which allows certain substances to pass through and keeps others out. Plants depend on osmosis. Minerals dissolved in water pass from the soil to the plant through root membranes. Osmotic pressure probably helps raise the sap to the high branches of trees.

OSWEGO TEA *(Monarda didyma)* MONARDA. This perennial, which grows to four feet under ideal conditions, has very aromatic foliage. Large, brilliant scarlet flowers from summer to early fall entice hummingbirds.

Oswego (so named by the Shakers who found the herb growing in profusion near Oswego, New York) was later known as bee balm. The healthful refreshment and delicious fragrance of this plant gives us a clue to its preference among native American teas as does its down-to-earth aesthetic value in the flower garden. Sage or basil, freshly ground, or dried peels of orange or lemon, give variety to the tisane.

OTAHEITE ORANGE *(Citrus limonia).* Because of its shape and color, the otaheite lemon is commonly known as the otaheite orange. The fruit is small to medium in size, round, orange-colored, pulp orange, juicy, blandly sweet. The plant is sometimes classed with limes but the purple flower buds and outer petal surfaces indicate a lemon relationship. This naturally small shrub, usually raised from cuttings and grown in pots, is very ornamental. It is fragrant when in flower and attractive in fruit.

OXYDENDRUM ARBOREUM. Sourwood, Lily of the Valley Tree. Closely rivaling the dogwood in interest and the beauty of its flowers and foliage, this plant is to summer what dogwood is to spring. The small, fragrant, bell-shaped flowers resemble lily of the valley and are borne during July and August in showy clusters eight to ten inches long. The attractive leaves assume deep red and scarlet tints during the autumn and form a contrast with the interesting seed pods. The plant is slow growing, which is considered an asset under certain conditions.

P

PANSY *(Violaceae).* The "flower with a face" is really a cultivated variety of violet. The lovely blossoms may be purple, violet, blue, yellow, white, brown, or a mixture of all these colors. In some climates pansies are best planted in the fall. The more you pick, the more pansies will bloom. If allowed to go to seed, they will stop blooming. You can increase pansies by taking cuttings from the center of the plant, but side shoots and branches will also grow. Place in a mixture of sand and loam, shade them from the sun, keep moist, and they will soon strike root.

Wild pansy *(Viola tricolor).* Germinates well if grown near rye. And the growth of rye is improved by a few pansies. The same is not true if pansies are grown with wheat.

Pansies should be picked and picked, and they will keep blooming.

PAPAW *(Asimina triloba).*   Powdered aerial portion has some effect on mealyworms.

PAPER FLOWER *(Psilostrophe tagetina).*   This very showy perennial covers itself with bright yellow flowers, forming a cushionlike mound. It is excellent as a border plant. Flowers dry and stay on the plant, making them valuable for use in dried arrangements.

PASQUE FLOWER *(Anemone patens)* WILD CROCUS.   This wildflower, one of the first of spring, has lovely, large, pale blue or violet, bell-shaped flowers. They are two to three inches in size and appear before the foliage.

   Sow outdoors in fall for germination the following spring, or cold stratify thirty days and sow in spring. The plants are nice in rock gardens and also for dried arrangements; the decorative seed heads look like fuzzy pom-poms. The name of the Pasque flower means Easter flower and is sometimes used to dye Easter eggs green.

In Russia the pasque flower is used to dye Easter eggs green. It's one of the earliest flowers of the spring.

PASSION FLOWER *(Passifloraceae)* MAYPOP. According to legend, early`Roman Catholic missionaries named these plants. They thought the ten colored petals represented the ten apostles present at the crucifixion. Inside the flower, colored filaments form a showy crown, which was thought to represent the crown of thorns. The five pollen-bearing anthers suggested Christ's wounds. The division of the pistil represented the nails of the cross. The bladelike leaf was symbolic of the spear that pierced His side. The coiling tendrils suggested whips and cords. Be this as it may, the flower is truly beautiful. The giant granadilla is red, violet, and white and is grown extensively in certain tropical countries for its fruits which are believed to be aphrodisiac.

PASTEURIZING SOIL. For pasteurizing potting soil to use for seed starting and houseplants, follow these directions: thirty minutes in a 140°F. oven is optimum for killing undesirable pests and will not destroy all beneficial plant organisms; thirty minutes at 180°F. kills most weed seeds and all plant pathogenic bacteria and fungi; temperature above 185°F. may damage the soil's chemical structure. Use a reliable oven thermometer. Soil should be moist, not wet. Spread evenly on an old cookie sheet or tray. Soil mixes should not be high in manure or compost because the heat will release too much fertilizer at one time and be toxic to plants. Pots and occasionally tools also need decontamination. A one-to-ten bleach solution will sterilize them.

PATCHOULI *(Pogostemon patchouly).* The fragrant oil of this shrubby East Indian mint is favored for perfumes.

PATIO PYRAMID PLANTER. A pyramid planter can be a most interesting focal point for your garden, balcony, or patio. A lot of different fruits and flowers can be shown off to advantage. Or use for a miniature herb garden. Even midget vegetables will grow in one of these. For indoors the planter is mounted on casters for easy movement; watering is provided for through a top reservoir.

PELLETED SEEDS. Pellets—and in the case of petunia seed they are about the size of shot—contain a little plant food as well as seed disinfectant. Easy to sow in flats, they often germinate and grow much better than common seeds. Seedlings suffer less from damping off. Pellets also allow you to space the seeds, thus eliminating the disagreeable job of thinning.

Head Start Transplants by Northrup King are practically foolproof for the gardener. Punch 'n Gro, the first seedling kit on the market, offers the head start seedling kit that waters itself. Water travels through a wick leading from a separate water receptacle into the growing medium. These are available in many flower and vegetable varieties.

PEPPER *(Capiscum)* Red Pepper, Chili Pepper. This tender annual or perennial from South America belongs to the potato family, Solanaceae. The solitary, white, star-shaped flowers are followed by juiceless berries or pods, variable in size and shape, at first green and, when ripe, yellow, red, or purple. Many of the newer types make attractive pot plants.

Capiscum has the same blood-stimulating effect as niacin, producing warmth. According to herbalist Jethro Kloss *(Back to Eden)*, "Capiscum is a pure stimulant . . . ultimately reaching every organ of the body. It creates a sensation of warmth, which afterwards becomes intense." Capiscum has been known for centuries as an aphrodisiac, due largely to its warming effect.

PEPPER JUICE. A California study found that natural juices squeezed from succulent plants such as green peppers were effective in protecting other plants from viruses. They work against diseases transmitted by insects or wind. The sprays were found effective against tobacco mosaic virus, potato virus, and several other viruses carried by aphids. Strangely, the compounds do not kill the viruses but change the plant so it is not susceptible.

Old-time gardeners planted hot peppers among their flowers to discourage insect pests. Of the ornamental peppers Midnight Special, with nearly black leaves and red fruits, is very lovely and unusual.

PEPPERMINT *(Mentha piperita)* Lamb Mint. The plant produces many creeping stolons (runners) which spread quickly in favorable environments. It is propagated by pieces of the runners and, like many of the mints, can be increased from slips (clones) planted in moist, sandy places.

Leaves are darker than those of the spearmint, larger and not as crinkly. The plant has a reddish tone. Even the small leaves near the rosy lavender flower spikes have red on their margins.

A medicinal oil is expressed from the dried leaves of flowering plants and is considered a powerful analgesic. The oil is also used to flavor tooth powders, pastes, and washes. Peppermint is used in cooking and for flavoring candy.

PERFUME. This invisible quality of flowers is one of their most important assets, however, many of our modern hybrids have lost their fragrance. The so-called "unimproved" kinds often retain this. The modest little, sweet-scented candytuft, *Iberis odorata*, is an example. The wild carnation seems to spray its admirers with its spicy incense. The scent of roses, lilac, and violets is enchanting. Most of the older varieties of iris are also sweet with perfume. If you would have fragrance in your flower garden, seek out the older, less showy varieties. Note also that flowers are less scented in periods of extreme heat and drought.

PERUVIAN GROUND CHERRY *(Nicandra)*. This is a pretty little plant with small pale blue flowers. It is also said to mean death to any bug partaking of its foliage. It grows in shade but produces more flowers in full sun. Set the seeds in fairly rich garden soil. Its blooming period (and its effectiveness) can be extended by keeping fading blossoms picked.

PESTS. Herbs and flowers that act as trouble-shooters for various pests include:

Pest	Plants
Ants (and the aphids they carry)	pennyroyal, spearmint, southernwood, tansy
Borer	garlic, tansy, onion
Cutworm	tansy
Eelworm	marigold (French and African)
Flea beetle	wormwood, mint, catnip
Fruit tree moth	southernwood
Gopher	castor bean
Japanese beetle	garlic, larkspur (poisonous to humans), tansy, rue, geranium (use white geranium)
Leafhopper	petunia, geranium
Mexican bean beetle	marigold, rosemary, summer savory, petunia
Mice	mint
Mole	spurge, castor beans, mole plant, squill
Nematode	marigold (African and French), salvia (scarlet sage), dahlia, calendula (pot marigold), asparagus
Plum curculio	garlic
Rabbit	allium family
Rose chafer	geranium, petunia, onion
Slug (snail)	prostrate rosemary, wormwood
Squash bug	tansy, nasturtium
Striped pumpkin beetle	nasturtium
Tomato hornworm	borage, marigold, opal basil
White fly	nasturtium, marigold, nicandra (Peruvian ground cherry)
Wireworm	white mustard, buckwheat, woad.

PETUNIA *(Petunia)*. This, one of the world's greatest summer plants, falls into four types: grandiflora doubles, grandiflora singles, multiflora doubles, multiflora singles. Petunias can "take it" but will bloom better and longer if fed liberally once a month—or if diluted fertilizer is added each time you water. Petunias live over the winter in mild climates. They also attract beautiful moths at night, and the fragrance of some kinds is very pleasing.

Never underestimate petunia power; they perform well everywhere in the garden. They thrive in pots, flower beds, greenhouses, or wherever they have a sunny location. And they can be grown from cuttings or from seeds. Petunias help to protect beans from the Mexican bean beetle.

PEYOTE. See Mescaline.

pH. Soil pH is important, whether in the flower or vegetable garden. Soil tests for acidity may be made at home using one of the gardeners' soil test kits; or a soil sample may be sent to a professional laboratory. Ask your county agricultural agent for information on your local soil testing facilities.

To get a reliable soil sample, use a clean spade to cut a one-inch thick, seven-inch deep slice of undisturbed soil. Take similar samples from several points in the garden; mix all in a clean bucket. Then dip out a pint jar of the mix to take to the laboratory, being sure to label it with your name and address. The lab will charge a small fee to cover costs of chemicals and a technician. The report will include not only the pH level, but usually the kind and amount of fertilizer recommended to bring the soil to a good balance of nutritional elements.

pH preferences of some commonly grown flowers are: chrysanthemum 5.7 to 7.5, hemerocallis 6.0 to 7.0, iris 5.0 to 6.5, ageratum 6.0 to 7.5, begonia 5.5 to 7.0, marigold 5.0 to 7.5. They will attain their best growth in this pH range and are less likely to do well above or below.

PHELLODENDRON *(Phellodendron)* Cork Tree. These are handsome, deciduous trees with short trunks and widely spreading branches. Male and female flowers are borne by different trees in summer.

Phellodendron belongs to the rue family, Rutaceae, and several kinds have the aromatic odor peculiar to other family members. They also share in the repellent properties of rue and a decoction made from the bark is repellent to insects.

The name is taken from the Greek, *phellos*, cork, and *dendron*, tree, and refers to the corky bark of several kinds. The Amur cork tree *(Phellodendron amurense)* is one of these kinds and is prevalent in Manchuria, northern China, Korea, and Japan.

PHILADELPHUS *(Philadelphus)* Mock Orange. Mostly hardy, these deciduous shrubs vary in size from small dense bushes two to three feet high, to large ones fifteen to twenty feet high and equally wide in diameter.

Breeders have produced many beautiful hybrids of mock orange. One of these hybrids, *Philadelphus virginalis*, is among the best and most fragrant. The sweet-scented mock orange, *P. coronarius*, is the commonest.

Its flowers are strongly scented, and although they are delightful in the garden, their scent is too strong indoors for many people. The double-flowered varieties are less strongly scented than the common kind and last longer as cut flowers.

Mock oranges are among the oldest shrubs in cultivation, dating as far back as the sixteenth century. Like the lilac they were one of the first shrubs brought to America and planted in the dooryards of the early settlers. The small, inedible fruits are often used as pomanders.

PHLOX *(Phlox drummondii).* The word phlox comes from the Greek word for flame. Its brilliantly colored flower, however, never becomes flame colored. Phlox are a true North American species and are favorite garden flowers because they are hardy and grow well in fertile soil. All annual phlox are derived from Drummond phlox, a species that grows wild in Texas. The familiar sweet William, whose bluish or pale lilac flowers are among the early summer blossoms, also belongs to the phlox group. Phlox are deliciously fragrant and for a long season in summer they dominate the garden.

Phlox are also attractive in hanging baskets with browalia or lobelia. They make a good ground cover with nicotiana or zinnia borders and are nice for beds and edgings. Sow where they are to grow; they dislike transplanting.

PIGGYBACK PLANT *(Tolmiea menziestii).* The piggyback has the fas-cinating habit of growing baby plants on top of the mature leaves. The other surprising fact is that it grows wild along the Pacific coast from northern California to Alaska. If you live where winter temperatures seldom fall below 10° F., plant piggyback outdoors. In a shady, moist rock garden, a single plant will soon multiply into a colony. Scattered about the floor of a woodland, piggyback looks lovely in the company of hardy ferns.

Indoors it is definitely a winner. Give it good light and keep it a bit on the cool side with a good soil mix consisting of equal parts all-purpose potting soil, sphagnum peat moss, and vermiculite. Don't let it stand in water but keep it evenly moist at all times.

PILLOWS. Flowers and herbs for scented pillows were once very popu-lar. The delicate fragrance of herbs or blossoms is released when pres-sure is put on the pillow. The herbs may be mixed with the pillow stuff-ing, used entirely as a stuffing, or put into sachet bags and placed inside the pillow. Scented pillows can be made from: calamus, lavender flow-ers, lemon verbena, meadowsweet, orris root, rosemary, rose geranium, sweet fern, or woodruff. Pillows stuffed with white pine needles are de-lightful.

PINCH A PLANT. Pinching Back, Pinching Out.   These terms have the same meaning and are used by gardeners to describe the removal of the growing tip of a shoot to ensure the development of side shoots. The term pinching out is also used to describe the complete removal of small side shoots, as is done when tomatoes are trained to a single stem and when the stems of chrysanthemums, fuchsias, and other plants are trained to form "trunks" of standard (tree-form) specimens.

Tip pinching makes a plant grow bushier; just pinch out the growing tip of the tallest stem close to a leaf joint.

The following annuals are improved by proper pinching: ageratum, antirrhinum, carnation, cosmos, phlox, salvia, tagetes, verbenas, dwarf-type dahlias (grown as annuals), and petunias.

PINE (various species).   Pine tar oil improves standard codling moth baits.

PLANTAIN *(Plantaginaceae).*   This weed is often troublesome to gardeners, its seeds being spread by birds which eagerly eat them. Yet it has value as an emergency measure to stop bleeding. Crush or bite the leaves to let out the juice; apply directly to the wound. Bleeding will stop, even from a deep cut.

Plantain has been used for hundreds of years for healing broken bones. Keep a few plants in the garden in case of need.

Add tender heart leaves in early spring to a green salad.

PLANTERS.   Want something different in a container? Gather a sculptured piece of driftwood from a lake's rim and fill it with lacy green ferns set with their own peaty earthballs in plastic bags, punctured at the base for drainage.

For indoors look for a fine piece of handcrafted pottery, a dainty basket, or a filigree of old iron; Western boots, well worn, also make unusual planters. An old coal scuttle is dandy for summer flowers. It does not tip over, can be punched with drainage holes, and there is lots of room for deep root growth.

Something extra special for an apartment balcony is a bird cage on its stand. Paint it black, fill it with a tumble of white marguerites, or red or yellow cascade petunias. Or paint it turquoise blue and use pink geraniums, white petunias, and green ivy.

Seashells make attractive planters. If you prefer not to deface a shell by cutting or drilling to provide drainage, water the plants sparingly. Plants such as velvet plant (Gynura), aloe vera, succulents, sedums, and miniature English ivy will grow well in these containers. See also Patio Pyramid Planter.

PLANT MIMICRY. The orchid *Trichoceros parviflorus* grows petals to imitate the female of a fly species so exactly that the male attempts to mate with it and in so doing pollinates the orchid.

The carrion lily develops the odor of rotting meat in areas where only flies abound. Other flowers which rely on the wind for pollination do not waste their time making themselves fragrant or beautiful to appeal to insects or birds but remain relatively unattractive.

PLANT NEWS YOU CAN USE. The color of house walls affects light intensity for indoor plants. Flat white, not glossy or semi-glossy, is the most efficient reflector of available light.

Do plants sleep? Plants sleep during a period called dormancy. They are affected by the cycles of winter, spring, summer, and autumn, as well as wet, dry, cold, and hot seasons. Many plants also close up and sleep at night.

Want instant garden color? Try annual bedding plants such as petunias, geraniums, salvia, ageratum, zinnias, and marigolds. For best effects, mass flowering annuals in large clumps of color rather than a staccato line of single plants.

Want more flowers? In midsummer, many perennial garden flowers begin a rest period. Divide at this time to form new plants. Suitable candidates include peony, German iris, Oriental poppy, madonna lily, painted daisy, phlox, and columbine. Crown division is one of the easiest methods. Usually plants that flower in spring and early summer may safely be divided in late summer and fall. Those flowering in summer and fall should be divided in early spring before new growth appears.

To make crown divisions, lift the plants carefully and remove some soil from the roots. Cut the crown into several pieces with a knife. Use the individual sections of vigorous plants to make new plants.

How often to divide? Peonies may remain in the same spot for many years. Divide shasta daisies and phlox every three years. Replant daylilies and iris every five years. Divide chrysanthemums and hardy asters every two to three years in the spring. Some plants, such as Oriental poppies, do not adjust well to moving; transplant these only when they lose vigor from overcrowding.

Don't throw away old pantyhose. Use them for tying up small sugar pumpkins or tomato plants to trellises or rings. Pantyhose is also good for cucumbers, vining squash, and melons. To protect cherries, cut the legs off old pantyhose and draw them very carefully over the limbs of the trees before the cherries ripen. This works well with a tree too big for a net, and will give perfect, unmolested cherries.

When air layering plants, place wet sphagnum moss or other rooting medium in old nylon hose to wrap around the prepared spot on the plant. Wrap again with a plastic bag and tie it down. The nylon hose keeps the rooting material in place and the roots grow right through it.

Why do some plants live and others do not? Like people, plants do not always respond in an exact relationship to their environment and care. Some have a will to survive, others don't.

An occasional tulip bulb planted upside down will circle laboriously around and reach up triumphantly for the light. A cactus growing in the desert will top itself with a brilliant blossom. Tiny seedlings seem to have a built-in will to live. And, of course, some varieties of plants have it more than others; for example, it's very hard to kill sansevieria (mother-in-law tongue) or a rubber plant. And some flowers and herbs grow and spread, seemingly on their own.

PLANT QUARANTINE.   Plant quarantine is a law that regulates the movement of plants and other materials that may carry a plant disease or insect pest. The quarantine keeps the disease or insect from spreading from infested areas to those free from these hazards. Some laws list plants which may not be shipped in or out of a locality. They may also give directions for moving, packing, and labeling.

In a quarantine, officials examine all plants at the border of the quarantined area and keep out the dangerous types. Foreign plant quarantines control the shipping of plants from other countries.

PLANTS PROTECT THEMSELVES.   Plants have a strong survival instinct and have developed various means of self-protection. Mechanical weapons include thorns on roses, prickles on thistles, and spines on cacti. Sumac (some types) uses poisonous chemical weapons, and nettles have irritating acids.

PLANTS WITH PARTY TRICKS.   Among the amusing flowers of the plant world is the mouse plant (*Arisarum proboscideum*). This useful, low-growing ground cover is related to the arum lily. The flower is in the form of a single spathe which by some freak, takes almost exactly the shape of the rounded hindquarters of a tiny mouse with a long curling tail. Hold it in your hand as if it were a mouse; friends will probably reward you with gratifying squeals.

Field-mouse-ear chickweed (*Cerastium arvense*) is a beautiful flowering native plant. Its name "mouse ear" comes from the shape of its leaves. The blossom is large, white, and star-shaped.

Silene armeria is one of the catchflies, related to the campions. The stems are divided into sections by nodes; below each is a dark, sticky patch. And they do actually catch flies and many other tiny insects.

Woad *(Isatis tinctoria)* is a little known plant which the ancient Britons used to paint themselves blue. Through a rather complicated process, blue dye can be made of it for dyeing cloth.

Gay feather *(Liatris pycnostachya)* is an upside down plant. Unlike other flowers that grow in a spike, this herbaceous perennial blooms from the top first, the opening buds gradually working their way down the stem.

An easy perennial to grow is *Gaura lindheimeri*, four feet high, which when in flower looks like a cloud of white butterflies. The flowers, which have pink buds, have only four petals. Each petal spreads out like the wings of a butterfly, while the long stamens resemble antennae.

Dictamnus albus, the gas plant, gives off a volatile gas which on a hot still day can be ignited by a match held near it and it will flare for a moment without the plant suffering any damage.

Obedient plant *(Physostegia)* demonstrates its obedience when you touch the flowers. They can be moved up, down, or sideways and will stay where they are put.

POCKETBOOK PLANT *(Calceolaria)*. This multicolor perennial is spotted orange, yellow, or red with blossoms which mimic miniature pocketbooks. The plant is very effective in mass beds or along a shaded lawn area.

POINSETTIA *(Euphorbiaceae)*. This plant of the spurge family has tiny flowers surrounded by large, colored *bracts*, or special leaves. The bracts are usually bright red but may be yellow or white. The brilliant red bracts contrast with the green leaves and make the poinsettia popular during the Christmas season. In tropical and subtropical regions, the poinsettia thrives outdoors. It may grow two to ten feet tall. It is a popular garden shrub in the southern states and California. In cold climates it must be grown indoors. As a potted plant it grows from one to four feet tall.

The mealybug is sometimes a problem. Alcohol is lethal to the mealybug. Dip a small stick wrapped with cotton in alcohol and touch it to a pest for just an instant.

Root aphids may cause your plant to become weak and stunted, and plants may die in severe cases. To make it difficult for these pests to get to the roots, pack the soil firmly around the plant.

Poinsettia scab is sometimes prevalent in summer. Prune and burn scab-infected branches as soon as noticed.

POLLEN. The tiny yellow grains seen in most flowers are pollen. They are used to form seeds. Plants make the pollen in the saclike anthers of their flowers. The anthers are the male organs of reproduction. The fe-

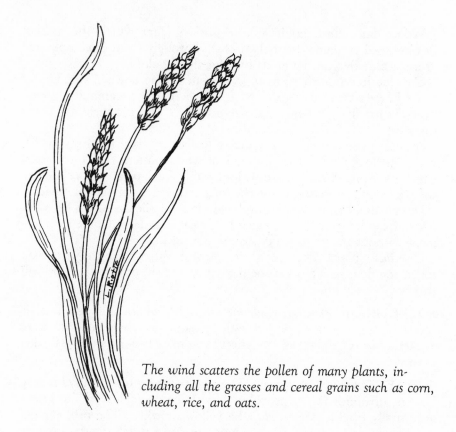

*The wind scatters the pollen of many plants, in-
cluding all the grasses and cereal grains such as corn,
wheat, rice, and oats.*

male organs include the pollen-receiving stigma leading to the ovary,
which is the egg-bearing part of the plant. Pollination is simply the
transfer of pollen from the anther to the stigma. When this occurs fertilization has taken place.

Self-pollination occurs in flowers that can transfer pollen from their
own anthers to their own stigmas. Cross-pollination means that the
flower must depend on wind, insects, birds, flies, or some other means
to carry its pollen from one flower to another.

Bee pollen has been considered a great energy producer for centuries
and it is reportedly consumed by many athletes. It is also believed to be
an aphrodisiac.

The pollen of most plants is highly inflammable. When thrown on a
red hot surface, it will ignite as quickly as gunpowder.

POLLINATION BY INSECTS.   Many insects are helpful as pollinators. Odor usually repels or attracts them, but color also plays a part.
Many insects do not see red (which attracts hummingbirds, for instance), but can see ultra violet, which we cannot see.

POMANDER. Take a small thin-skinned orange and press whole cloves into it until the surface is entirely studded. Roll the orange in powdered orris root and powdered cinnamon, patting on as much as possible. Wrap in tissue paper and place in a dry, well-ventilated area for several weeks. Remove paper, shake off surplus powder, and the pomander is ready for use. Hang by a ribbon in a closet where it will share its fragrance and aroma for many weeks. The mock orange (*Philadelphus virginalis*) is also used for this purpose.

POND, PLANTING AT EDGE.   Suitable plants for the edge of a pond are: *Iris pseudacorus*, *Iris versicolor*, arrowheads, sweet flag, cardinal flower, flowering rush, purple loosestrife, marsh marigolds, and astilbes. Tiger lilies do not like wet soil, but *Lilium canadense* and *Lilium superbum* do.

POOR-MAN'S- WEATHERGLASS (*Anagallis arvensis*) PIMPERNEL.   The small starlike, bright blue flowers of this low-spreading annual close with the coming of bad weather—believed by many to be a sure sign of rain. Plant seed in spring in full sun and poor, sandy soil. The plant will flower from May to August.

The water lily is seen in a variety of colors. Small air sacs in the leaves keep this plant afloat.

POOR SOIL WORK. Most plants fail because they are improperly planted. Far too many of us place a $10 plant in a $1 hole. Prepare the best possible foundation for your plant. If the plant is bare root, make the hole large enough for the roots to spread out naturally.

Plant purchases are often made on impulse with no thought to where the plant will be placed, nor what its ultimate size will be, nor how it will adjust to other plants in the immediate vicinity. The urge to use plants that will give a quick effect often results in overcrowding.

Flower seeds often are sown too thickly. Pull out or transplant the surplus plants as soon as they begin to crowd each other.

POPPY *(Papaveraceae).* Many different flowers are popularly called poppies, some native, some introduced. In this flower family are: the California poppy *(Eschscholzia californica),* corn or Flanders poppy *(Papaver rhoeas),* Iceland poppy *(Papaver nudicaule),* matilija poppy *(Romneya coulteri),* Mexican gold poppy *(Eschscholzia mexicana),* and the yellow poppy *(Papaver radicatum).*

Opium comes from the young capsule of the poppy where the seeds develop. To obtain it, workers slit the capsules late in the day. The milky juice that seeps out solidifies overnight, and is collected by hand the next day. It takes about 120,000 seed capsules to yield twenty-five to forty pounds of opium.

The poppy has a bad name, because opium is made from it. Many varieties look well in groups in the perennial border or in rock gardens.

Poppies are actually robbers of the soil and inhibit the growth of winter wheat. They dislike barley but will lie dormant until winter wheat is sown in the field.

PORTULACA *(Portulaca grandiflora)* Rose Moss. This most gaudy of coverings for very dry spots is first cousin to the weed pusley (or purslane). It grows and does well in hot, dry, shallow soil where no other flower will; for seaside gardens it is indispensible. Portulaca grows six to eight inches high and is of trailing habit. The blossoms are red, magenta, orange, and white, appearing from July to October. Culture is simple. Just scatter the seeds over the surface of raked ground when the weather is warm.

POTATO CHIP CAN. Those tall potato chip canisters can be used as dusters for flowers and vegetables. Punch holes in the bottom and drop in a few marbles or pebbles for agitators; fill the canister two-thirds full of rotenone, dipel, or other garden insect dust; put on the lid; and you are ready to do battle.

POT MARIGOLD *(Calendula).* This annual with yellow or orange blossoms is a nice background flower for pansies and candytuft. Plant it in fall for color through winter and spring. Pot marigold is good against asparagus beetles, tomato worms, and many other insects.

Portulaca is a delightful plant for carpeting the ground. It thrives in sandy soils in sun and heat.

The pot marigold, or calendula, has been a popular annual for centuries. In the Tudor period it was known as the Sunne's hearbe, *or the* Sunne's bride, *and the name marigold is linked with the Virgin Mary.*

POTPOURRI *(Provence)*.   There are many recipes for potpourri. This is one of the most delightful. Gather as many as possible of the following kinds of scented flowers: petals of the pale red and dark red roses, moss roses and damask roses, and acacia; and the heads of pinks, violets, lily-of-the-valley, lilacs (blue and white), orange blossom and lemon blossom, mignonette, heliotrope, narcissi, and jonquils; with a small proportion of the flowers of balm, rosemary, thyme, and myrtle. Spread materials out to dry. As they become fully dry in turn, put them into a tall glass jar, with alternate layers of coarse salt (uniodized) mixed with powdered orris root (use two parts of salt to one of orris). Pack the flowers and salt-orris until the jar is filled. Close the jar for one month, then stir all up and moisten with sufficient rose-water to penetrate to the lowest layer. Cap with a muslin cloth tightly tied, and use in cotton bags when wanted to scent drawers, linen closets, and clothes hangers.

PRIMROSE *(Primula parryi)*.   In summer this flower has intense cerise flowers with a yellow eye. The plant likes full sun or partial shade, and wet feet! It grows well with candytuft, pansies, calendulas, and violas.

PROLIFERATION. In the plant kingdom proliferation means new growth by cell division or buds. Examples of this are the daylily which not only makes seed and may be divided by its tubers but also makes new plants at joints of the flower stalk, sometimes growing aerial roots. (See Daylily.)

The so-called airplane plants also propagate themselves in a similar manner by sending out runners.

PRUNING PRINCIPLES. Early flowering ornamental trees and shrubs form their buds in summer and fall. Therefore do any necessary pruning only during the month after they have flowered; if you prune them in late winter or early spring, you will be cutting off the buds. Some early flowering trees and shrubs are dogwood, crabapple, forsythia, rhododendron, rose, and viburnum.

PURSLANE. Purslane is a weed. Do not put it in your compost heap; it will survive and again be planted in your garden.

PUSSY WILLOW (*Salix mutabilis Discolor*). This one has dainty, pearly catkins. Cut for indoor decoration in January and February. Place them in water and watch them unfold. Children find these delightful. The French pussy willow (*S. caprea*) produces silver pink catkins that are deliciously honey-scented.

PYRETHRUM (*Chrysanthemum coccineum*). This interesting perennial grows to two feet with red, pink, or white daisies of three inches across. Crumble the flowers, fresh or dried, into water (or use a blender). Use the spray to control aphids and other soft-bodied insects. It is organic, safe, and easy, being the least toxic (to man, animals, and plants) of any insecticide available. It is also useful as a powder to control insect pests on pets or farm animals.

PYROSTEGIA (*Pyrostegia ignea*). This showy, tender, climbing shrub from Brazil produces rich crimson orange, tubular flowers in large drooping panicles. The name pyrostegia is derived from *pyr*, fire, and *stega*, roof, and refers to the upper lip of the flower. This high climber is a marvelous choice for covering the rafters of a large greenhouse or for growing outdoors on arbors in the South.

Q

QUEEN ANNE'S LACE (*Umbelliferae*). This delicately beautiful weed can spread and become a great pest. It is really a carrot that grows wild, but the roots are not edible. It may be either an annual or a biennial.

A far better substitute is bishop's weed (*Ammi majus*). This lovely 2½-foot annual has lacy, white flowers like Queen Anne's lace and is widely cultivated for cut flowers. It will grow just about anywhere.

Queen Anne's-lace is a hardy weed related to the domestic carrot. The roots are not edible.

QUEEN OF HUNGARY'S WATER.   This tonic lotion was once made and used by ladies everywhere. Not only was it used for perfumery but a few drops taken internally were recommended for nervous ailments and mental depression. Here is the recipe: 2 tablespoons dried rosemary flowers, 1 grated nutmeg, 2 teaspoons cinnamon, and 1 tablespoon sweet cicily leaves (if available). Pulverize all the ingredients (a mortar and pestle is helpful) and mix well together. Add a quart of pure alcohol. Let the mixture steep for ten days. Then strain off and bottle. Apply on cloths wrung out in cold water, and place over forehead to allay headaches and soothe fevers. With fevers also apply to the pulse of the wrists. (See COSMETICS.)

QUINCE *(Rosaceae).*   The quince, a shrub or small tree, is one of the loveliest members of the rose family. Though grown mainly for the fruit, the rosy flowers which bloom early in the spring are very attractive. They are long-lasting when cut. Placed in a low bowl, the cut flowers give a delightful oriental effect.

The pear-shaped fruit has a golden-yellow color and a fragrant smell. Quince is never eaten fresh as it is quite hard and has an acid taste, but it is very pleasing when cooked or used in marmalades, preserves, and jelly. Plant quince shrubs with garlic; it improves the flavor of the fruit.

The quince tree produces fruit used to make a delicious jam. The tree has long been cultivated, but has never been popular in this country.

R

RABBITS, COTTONTAILS.   Their name comes from the fluffy, snow-white underside of the tail. They like to hide in heavy thickets or dense grass. Though often seen in daytime, they usually come out at night to gather their food. When there are too many of them, they can become serious pests by eating growing hay, vegetables, grapevines, and young fruit trees. And they also like to eat young flower plants. If you find your plants being eaten and no trace of insects, cottontails may be infesting your garden by night.

Onions, also a thin line of dried blood, or blood meal sprinkled around the edges of the garden may discourage them. Or try powdered aloes, wood ashes, ground limestone, or cayenne pepper.

RAFFLESIA.   This small genus of plants has huge flowers but no leaves or stems. The flowers grow as parasites on the stems and roots of several cissus shrubs in Malaya. One species of rafflesia produces flowers more than three feet wide and weighing up to fifteen pounds. The stamens and pistils grow on separate flowers, and need some agent to pollinate them. The flowers have five wide, fleshy lobes and an unpleasant odor.

RANUNCULUS (*Ranunculus*).   This lovely perennial in yellow, orange, red, white, and pink is sensational in masses. Or use with snapdragons, pansies, and daffodils.

Rabbits, cute as they are, can level both flower and vegetable gardens during a one-night visit. They're hard to discourage, but try one or some of these: dried blood, blood meal, cayenne pepper, or wood ashes.

RASPBERRY *(Rubus idaeus).* Few realize just how beneficial this wild variety of the rose family may be. Flowers are of true rose-form, pure white, with prominent stamens. Stems are ranging and thorny; fruits are a brilliant red, darkening when ripe and juicy.

The foliage of raspberry plants has a very active principle, *fragrine*, which has a special influence on the female organs of reproduction, according to Juliette de Bairacli Levy in her book *Herbal Handbook for Everyone.*

Raspberry tea, made by placing one ounce of dried leaves into 1½ pints of water and simmering for twenty minutes, is helpful as a tonic during pregnancy. It is said to relieve morning sickness. Both foliage and fruits are considered an aid to easy childbirth and contributory to the health of both mother and child. The plant, acclaimed as a remedy for sterility, is also thought to be helpful against both frigidity and barrenness. Cultivated raspberry plants retain some of the fragrine principle but the wild plants are believed to be most potent. Wild plants are not subject to mosaic.

RATTLEBOX *(Crotalaria)*. Prized not only for its racemes of yellow sweet-pea-like flowers, it also is toxic to nematodes.

Nematodes, worm like inhabitors of the soil, are microscopic in size, and the root knot types injure many garden plants. These tiny worms like sandy, warm soils and have been known to devastate entire gardens. Chemical soil disinfectants kill them but kill the plants as well, so the discovery of a protective flower is very valuable. The pretty crotalaria, *C. retusa*, grows three feet high with maroon reverse petals. *C. spectabilis*, the species tested for nematode control, is four to five feet tall with smaller flowers than *retusa*. From its seed pods, which rattle when shaken, crotalaria derives its common name.

Crotalarias have value also as legumes; they can be turned under to add nitrogen to the soil. However, because they are poisonous to livestock, confine them to the garden.

REDBUD *(Cercis)*. The redbud and the dogwood come into blossom at approximately the same time and complement each other, one rose pink and the other sparkling white. Members of the bean family, redbuds are also nitrogen-fixing trees.

RESURRECTION PLANT *(Cruciferae anastatica)* ROSE OF JERICHO. This strange plant lives in the deserts of Mexico, Palestine, and Syria. Actually a seed, its small branches are folded up into a ball. When ripened, it drops off the mother plant and rolls about in the desert winds until the rainy season. If placed in a saucer of water, the seemingly lifeless ball will open into a beautiful rosette of fernlike leaves.

Another resurrection plant is bird's-nest moss. This plant reproduces by means of spores.

ROCK GARDENS. There are hillside rock gardens, valley rock gardens, as well as those on level sites; there are rock gardens in sun and rock gardens in shade, as well as those that include water.

All rock gardens should include dwarf material in trees and shrubs. Nurseries that specialize in choice rock-garden plants usually have a good selection. The extremely dwarf evergreen trees, for example, grow slowly and never develop out of proportion to a miniature landscape.

Make the rock garden attractive in winter as well, by using a number of evergreen rock plants (many of which are technically shrubby or subshrubby), such as perennial candytuft and *Iberis sempervirens*. The rock roses *(Helianthemum)* are brilliant in early summer with myriads of satin-textured blooms resembling little wild roses in pink, crimson, orange, gold, white, and scarlet. The silver saxifrages are very useful for creating the all-the-year-round picture of true alpine rock gardens. For cooler, shadier rock gardens try the mossy saxifrages. The genus *Dianthus* includes many splendid plants well adapted for sunny locations and well-drained sites. There are also bellflowers or campanulas, thymes, stone-crops or sedums, primroses or primulas, houseleeks or sempervivums, gentians, bulbs, and ground covers.

ROOF TERRACE, PLANTS FOR.   Consider sun and shade just as if you were planting on the ground. And study the prevailing wind direction before you make plant selections and decide on locations. Given good soil and some protection, you should be successful with begonias, lobelias, nicotianas, and impatiens. Japanese yews make nice accent plants.

ROOTS.   Roots are one of the three organs that most plants must have in order to grow. The other two organs are the stems and leaves. The roots of most plants grow in the ground and draw their food material from the soil. But some plants have their roots in water or even in air.

Roots hold the plant in place and supply it with water and nourishing salts from the soil. Roots that form first and grow directly from the stem, are called *primary roots*. Branches of the primary roots are called *secondary* roots, and branches of these are *tertiary roots*.

Roots with different forms have special names. A primary root that grows much larger than any of its branches is called a *taproot*. When taproots grow very thick and store up food for the rest of the plant, they are called *fleshy* roots. A cluster of thick primary roots is called *fascicled* roots. Threadlike roots are *fibrous*. Roots may also grow on the stem or in other unusual places. These are called *adventitious* roots.

Roots are also called *soil roots*, *aerial* or *air* roots, or *water roots*, depending on where they grow. Roots that get their food from other plants are called *parasitic roots*.

ROSE, WILD *(Rosa)*.   Different species of wild rose are native to every state of the union except Hawaii. Often they represent the so-called transition shrubs between forest and meadow or prairie. Many sucker freely and can be invasive, but they are much hardier and more disease resistant than their hybrid relatives.

ROSE CHAFER.   The rose chafer is repelled by geranium, petunia, and plants of the onion family.

ROSE FACIAL MASK.   This treatment is especially helpful for oily skins. The ingredients are: ⅔ cup finely ground oatmeal, 6 teaspoons honey, 2 teaspoons rose water. Blend oatmeal and honey until well mixed. If desired, add more honey to make a smooth paste, blending it with rose water. Spread over clean face and neck. Leave on for 30 minutes. Best results will be achieved if you lie down and relax. Using a soft wash cloth and warm water, remove and follow with cold water or astringent.

ROSE FAMILY.   The rose family is one of the most important in the plant kingdom and includes about 2,000 species of trees, shrubs, and herbs. Some of the loveliest flowers and most valuable fruits belong to it. A few members of the family are: apple, apricot, blackberry, bridal wreath, cherry, cinquefoil, eglantine, peach, pear, plum, quince, raspberry, spiraea, and strawberry. Its many ornamental plants include the meadowsweet, mountain ash, and hawthorne. Plants of this family give us many useful products such as attar, an oil from rose petals, used to make toilet water and perfume. Several fine woods are used in cabinetmaking.

Plants of the rose family have regular flowers. Each has five petals, a calyx with five lobes, many stamens, and one or more carpels. They bear seeds, so they are classed as angiosperms. The sprouts have two seed leaves, therefore they belong to the dicotyledonus plants.

ROSE HIPS.   Rose hips have long been valued for their vitamin C content, but they are rich in vitamin E as well. The *Rosa rugosa* in particular is a special favorite for the various purées, jams, and syrups made from rose hips.

The American Rose Society suggests these planting pointers for roses: plant them where they'll get sun at least half the day. Plant during winter where ground isn't frozen. A raised bed works well for roses in many areas.

ROSEMARY *(Labiatae, Rosmarinus officinalis).*   This evergreen shrub of the mint family is noted for its fragrant leaves. It has tiny, pale blue flowers and dark green leaves. In masses, blossoming rosemary looks like blue gray mist blown over the meadows from the sea. Its name comes from the Latin *rosmarinus,* meaning sea dew. A thick growth planted around the flowerbed of prostrate rosemary will act as a border for snails and slugs; the sharp foliage apparently hurts their soft, slimy skins.

Cooks use the plant in seasoning and its oil is used in perfume. The oil is secured by distilling the leafy tips and leaves. It gives the characteristic note to Hungary water; eau de cologne cannot be made without it.

Rosemary oil is in all pharmacopoeias. The flowers are a stimulant, antispasmodic, emenagogue, and rubefacient, while the leaves are rubefacient (external application causing redness of the skin) and carminative (cleansing).

ROSE OF SHARON *(Hibiscus)* SHRUBBY ALTHEA.   The hibiscus shrub is distinguished by rose, purple, white, or blue flowers about three inches wide. The flowers appear in September when few other shrubs are in bloom. It does well even under unfavorable conditions, either in the city or the country and is a good shrub for the gardener who has little time.

ROSES.   Roses do not like boxwood because its outspreading, woody roots interfere with the roots of rose bushes. But garlic, onions, and other members of the onion family, including ornamental alliums, are beneficial.

ROSE, ATTAR OF.   A tiny, one-ounce copper vial of greenish yellow fluid—the essence of roses used in the world's most expensive perfumes—is more valuable than gold! The precious liquid concentrate (it takes over 100,000 roses to produce just one ounce) comes from the unique Valley of the Roses, where soil and climate have combined to make the finest rose scent in the world.

"Treated like a magic potion," according to Cyril Williams, a British perfume expert, "it's kept locked in bank vaults and fireproof, temperature-controlled safes. Small containers are insured for thousands of dollars."

The rose grown in the Kazanlik Valley, Bulgaria, for the purpose of making attar of roses, is *Rosa damascena trigintipetala.* It grows three to four feet tall and has one annual flowering. It has semi-double, rose red, 3½-inch flowers with stiff yellow stamens. This rose also makes wonderful potpourri.

The Turks brought these roses from Damascus when they conquered Bulgaria nearly 600 years ago. When they departed in 1878, they left them behind. "From a distance the fields look like a pink sea," said Valentine Ruskov, a Bulgarian trade attaché in London. "The Turks used to take baths in rose-scented water and before long the distillation of attar became a cottage industry." Ruskov explained that the roses are harvested in May and June before daylight, "so the sun doesn't dry the petals, which are boiled to remove the essence." He went on to say, "Other countries have tried to duplicate the essence, but have not been successful."

ROSES, CLIMBING.   The various types of climbers behave quite differently. Here are some good climbers which have stood the test of time: blaze, blossomtime, Chevy Chase, coral dawn, Dorothy Perkins, high noon, inspiration, Marechal Niel, mermaid, new dawn, and parade.

Support for climbers can be of conventional patterns, or design and build a support that is suited to your own needs. Roses must have a support to look their dramatic best.

ROSE SOAP.   Save odds and ends of leftover hand soap and grate or cut up finely. Add hot water that has had rose oil (about six drops) added to it, using enough to cover. Place over a low flame until soap dissolves. Pour into a clean cream carton to a depth of one inch and set aside to harden.

ROSES, UNIQUE.   Startlingly "new" roses appear from time to time; three of unique beauty and coloring are offered by Lakeland Nurseries.

Midnight Magic looks like the velvet black of midnight, so deep is the magenta coloring. The double blooms start in spring six weeks after planting and continue until frost. The nearly sable buds unfold to a breathtaking dusky-red rose, five to six inches across.

Green Dynasty is a startling green rose, a descendant of centuries of Chinese rose culture. This rarity starts blooming long before most roses realize it's spring. The silvery green buds unfold into the most exquisite mint green rose. The delicate four-inch blooms continue until frost and have a hauntingly sweet fragrance.

Blue Wedgwood is another miracle of nature. In its shapely foliage nestle clusters of pearlized lilac-tinted buds. These open into perfect flowers up to 5½ inches across, turning the cluster into a lavender blue symphony of beauty. The lilac and blue flowers are deliciously scented and bloom over a long season.

ROSES, UNUSUAL.   There are species of roses that botanists have discovered in various parts of the world and brought into cultivation. These roses are fascinating individualists for all developed distinct characteristics enabling them to survive in their native habitats. Some are

Miniature roses are being used to create exciting edgings, handsome beds, and, when potted, points of emphasis on patios. Give them a rich loam, and mulch plans in the winter with soil or straw.

extremely hardy and have one annual flowering; others, native to the subtropics, are tender and bloom repeatedly.

Rosa rubrifolia is a beautiful shrub, so named because everything about it is red, from the soft pink 1½-inch single flowers with their reddish brown calyxes, to reddish brown canes, dark greenish red foliage, and bright hips the color of Queen Anne cherries. Blooming off and on through the season, it grows to 6 feet, is quite hardy, and is native to the mountains of China and southern Europe.

Rosa roxburghii is more commonly known as chestnut rose, for the unopened buds look like little chestnut burrs. It is one of the most beautiful and unique roses in existence, with light green foliage whose new tips are colorful with shades of copper and gold, silvery gray branches that shed their bark as many trees do, and very double 2½- to 3-inch flowers of many small petals, glowing pink at the center with silvery pink on the outside. It blooms repeatedly and makes an excellent bank cover. It grows quite large in mild climates.

Rosa souliena was discovered in western China. Its relaxed canes lie on the ground and grow from 12 to 20 feet long. In June and July they are covered with corymbs of 1½-inch, single white flowers with a distinct fragrance that perfumes the air. It makes an excellent bank cover, for over the years its canes will take root and hold the soil. Grown on a retaining wall it will cascade to the ground like a waterfall. Growing up a tree such as a weeping willow the canes will fountain down very dramatically. When the petals fall, small orange hips form. Occasional flowers appear after its mass summer blooming.

ROTENONE. DERRIS.   Rotenone is an insecticide derived from certain tropical plants such as derris, cube barbasco, timbo, and several others. This contact and stomach poison is often mixed with pyrethrum. It is of low toxicity to man and animals. Like pyrethrum, it can be obtained in pure state only from pet shops and veterinarians. Devil's shoestring (*Tephrosia virginiana*) is the only native plant that contains rotenone. This is a common weed in the eastern and southern states, and its roots may contain as high as 5 percent rotenone. Rotenone is safely used on all crops and ornamentals but the period of protection is short.

RUBBER PLANT (*Ficus*).   This common houseplant is related to the fig. It grows well in house heat and lack of humidity, growing tall rapidly and living a long time. It grows even better if the pot is rich in minerals and the plant is given enough sunlight, water, and room. Place it outdoors in summer so it will get enough sunlight to last during the winter months. Should it be attacked by scale insects, spray the plant with nicotine. Commercial rubber does not come from these rubber plants, but from a tropical tree that belongs to the castor bean family.

RYANIA.   This plant-derived insecticide was discovered in 1943. It is a powder made by grinding up the roots of the South American plant, *Ryania speciosa*. Ryania has little effect on warm-blooded organisms but is useful in controlling corn borers, cranberry fruitworm, codling moth, Oriental fruit moth, cotton boll worm, and other insects.

While ryania may not reduce the number of harmful insects present, it protects the crop by making the pests sick enough to lose their appetites. Some species are not killed outright by it but are induced into a state of "flaccid paralysis." Ryania is recommended when there is an unusually large insect infestation and the gardener feels that nature needs a helping hand.

S

SAGE (*Salvia*).   These aromatic shrubs of the mint family have large, showy flowers. *Salvia officinalis* is the source of the spice. These plants require only the simplest of care; the salvias are the "purple sage" so often referred to in songs and stories of the Old West.

Sage is a delicious flavoring for sausage, pork, duck, and poultry dressing. A favorite herb of Marcus Aurelius, it has figured frequently in love potions, recipes, brews, and stews throughout the ages.

SANDY SOIL. Sandy soil, sometimes found near the seashore, need not be a problem if the right flowers are planted. Among annuals adaptable for sandy soils are calendulas, California poppies, sweet alyssum, marigolds, nasturtiums, portulaca, cleomes, and petunias. Shrubs may be Russian olive, *Rosa rugosa*, hydrangeas, rose acacia, tamarix, and Siberian pea tree.

SARSAPARILLA *(Smilax).* This group of woody or herbaceous vines has hardy, tuberous roots and veined evergreen leaves. The vines grow in temperate and tropical climates and bear small clusters of red, blue, or black berries. Some species yield the drug sarsaparilla, which was once widely used as a spring tonic. It is also used as a flavoring for soft drinks and medicines.

True sarsaparilla *(S. officinalis)* is related to asparagus. Mexican Indians have long used the roots of the vine in a concoction they believe cures impotence. In 1939 it was discovered that the sarsaparilla root contains large amounts of testosterone, and today much of the male sex hormone manufactured commercially comes from the plant.

The tea made from one or two teaspoons of the root is simmered slowly in a pint of water for five minutes and sipped slowly, the liquid held in the mouth for a time so that it can be absorbed into the buccal mucous lining of the mouth and better assimilated. (Indiana Botanic Gardens)

SCATTER PLANTS. Instead of using herbs alone in a stiffly formal herb garden, try scattering them throughout your flower garden. For their protective qualities include: marjoram, oregano, lavender, santolina, blessed thistle, camomile, lovage, chervil, lemon balm, and bergamot.

SCIENTIFIC NAMES. These are the names botanists give to all plants. They are in Latin and are the same in all countries. Scientific names have two parts, the genus and the species. For example, the prairie rose is *Rosa setigera*. This means that it belongs to the genus *Rosa* and the species *setigera*. The relationship of plants is determined mostly by their reproductive parts.

SEA ONION *(Urginea maritima).* These are tender bulb plants from the Mediterannean region, and tropical and South Africa. Tiny new plants grow on the "mother" under a thin skin. From time to time they pop out and take root as they fall off. They are easily propagated by the gardener if the small bulbs are picked off or allowed to drop and set in moist soil. Urginea belongs to the lily family, and its name is derived from the name of an Arabic tribe in Algeria, who were the first to use the bulbs medicinally. Syrup of Squills, a medicine, is produced from the bulbs, which are also good against rats.

SEDUM (*Sedum*). These succulent plants have attractive foliage and showy flower heads borne in late summer and fall. Plant in pottery containers, old shoes, boots, driftwood, or other unusual "pots" for a conversation piece, as well as in the rock garden. They are absolutely maintenance- and trouble-free.

SEED CATALOGS. Everything in a seed catalog is superb, magnificent, and usually resistant to something. But, just as in an insurance policy, you've got to watch the wording.

There's that phrase, "reseeds itself." That one you can believe, especially if it refers to something like bluebells-of-Scotland.

"Likes full sun" is also to be believed.

"Grows in shade" needs a little interpretation. What it actually means is "filtered sunlight," a spot dappled with a few shadows now and then, as under a tree. Few plants grow well in full shade.

"Naturalizes well" means that if you give the plant half a chance, it will take over. So beware.

"Grow in clumps of three or more" means to do just that. Grown alone it will barely make an imprint on your consciousness, yet grown in a small group the plant can be very lovely.

"May be divided" means you must—if you don't want the plants to choke themselves to death.

"Once established" means that getting this plant to do what you want may take a long, long, long time.

SEEDS. Almost all the plants mentioned in this book appear in the catalogs listed under Sources of Supply (see pp. 225). You can buy seeds with confidence from these dealers.

SEEDS, NUTRITIONAL QUALITIES. Seeds have been utilized by mankind for food since the beginning of time, but there are some that have very special qualities such as protein, vitamins, minerals, and hormones. Pumpkin seeds preserve the prostate gland and thereby, also, male potency. In Hungary, Bulgaria, and Turkey, among other countries, people eat pumpkin seeds as a source of vigor. Pumpkin seeds are rich in vitamin A and phosphorus, and several new varieties (Lady Godiva, for example) have seeds without a hard outer coat.

Bean seeds of all kinds, but especially broad beans, are also invigorating. Beans are rich in iron, copper, and phosphorus. Celery seeds also have been long celebrated for their stimulant powers. The soybean is rich in food values. It has more protein than beef, more calcium than milk, and more of lecithin than eggs. It is also rich in vitamins, minerals, and acids.

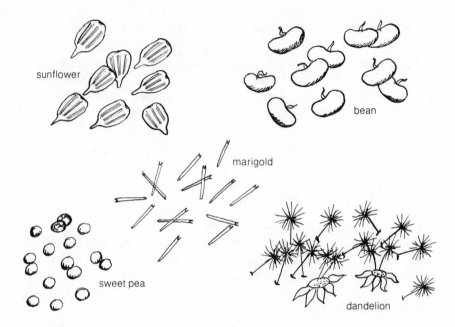

*A seed is a little bundle of determination-to-grow, and it is the most impor-
tant part of the plant. The roots, the leaves, the flowers all exist so there
can be seeds. As shown above, there's a great variety among seeds, from the
airborne dandelion seeds to the relatively large beans.*

SEEDS, SOWING. To evenly distribute tiny seeds, such as portulaca,
place them in a clean squeeze bottle such as mustard comes in. Turn
bottle upside down, press tip gently against soil where you want to seed
and squeeze lightly. A few seeds will come out, and when bottle is lifted,
soil covers them.

To make it easier to see small seeds shaken into a seedbed or furrow,
place them in a clean salt shaker with enough talcum powder to coat
them. This method also saves seed as you get more even distribution
and less need for thinning.

Here's a method to plant petunia seed directly into Jiffy-Sevens without failure. Take a wet pencil tip, pick up one tiny seed at a time, and
apply it to the moistened peat of the Jiffy-Seven. For insurance put two
seeds per pellet and remove one when both sprout.

SEEDS, SPROUTING. When starting seeds in pots use a plastic coffee
can lid to cover the pot until seeds sprout. It works better than fragile
plastic wrap to hold in moisture and also prevents quick cooling or heating when weather is changeable.

Two kinds of flower seeds sprout faster if they are tortured. One tough nut is the flower named canna. It is also called Indian Shot because the seeds of canna are round, heavy, and oily like buckshot. Nature made canna seed coats so that not all of them would sprout the first year. You can fool Mother Nature by cutting through canna seed coats to admit water. Use a triangular file or nail clippers. Soak the nicked seeds in warm water overnight, blot dry, and plant in warm soil; the seeds should sprout in two to three weeks.

Scald the seeds of the hibiscus called mallow. Bring water to a boil, turn off the heat, drop in the hibiscus seeds and leave them in the water overnight. The hot water won't kill the seeds, but it will cut through the natural oils.

SEEDS, STORING.   Until recently, home gardeners could do little to prolong the life of leftover garden seeds, especially those naturally short-lived such as onion, parsnip, delphinium, and larkspur.

Dr. James Harrington of the University of California, Davis, has developed an inexpensive method of storing leftover garden seed from open pollinated plants. (Saving hybrid seed is not recommended as they seldom come true.) He proposes a simple approach to dry seeds while keeping them cool; powdered dried milk is used as the dryer or "desiccant." Dr. Harrington recommends:

1. Unfold and lay out a stack of four facial tissues.
2. Place two heaping tablespoons of powdered milk on one corner. The milk must be from a freshly opened pouch or box to guarantee dryness.
3. Fold and roll the facial tissue to make a small pouch. Secure with tape or rubber band. The tissue will prevent the milk from sifting out and will prevent seed packets from touching the moist desiccant.
4. Place the pouch in a wide-mouth jar and immediately drop in packets of leftover seeds.
5. Seal the jar tightly using a rubber ring to exclude moist air.
6. Store the jar in the refrigerator, not the freezer.
7. Use seeds as soon as possible. Discard and replace the desiccant once or twice yearly. Dried milk is hygroscopic and will quickly soak up moisture from the air when you open the jar. Therefore, work quickly when you remove seed packets; recap the jar without delay.

This method is of special value to gardeners who schedule sequential plantings of short rows to keep a constant supply of flowers or vegetables.

SEEDS, SURVIVAL OF.   Seeds possess an enormous potential for survival. It is not unusual for large seeds, 100 years old and more, to grow when planted. During World War II in England, the bomb craters blossomed with flowers and other plants that had not been seen within the memory of man. Evidently the seeds were sleeping deep in the earth until brought again to the surface where light and moisture caused them to germinate.

In 1982 scientists tested lotus seeds found in an Asian lake bed deposit. They were found to be viable, and radiocarbon dating showed that they formed on lotus plants between 1410 and 1640. They are believed to be the oldest seeds ever to be proved still alive after long inactivity, and may be the world's oldest living tissues.

Other claims for this are doubted by scientists. These include stories of viable wheat found in the tombs of Egyptian pharaohs, and Arctic lupin seeds found in the Yukon and believed to be at least 10,000 years old.

SERPENTARIA *(Aristolochia serpentaria)* Virginia Snake-root. This small, aromatic perennial herb grows to a height of eight to fifteen inches. The flowers are usually hidden beneath the dry leaves and loose top-mould. Serpentaria is a remedy for snake bites.

SESAME *(Sesamum indicum)*. The tiny but exquisite flower may be pink or white. This plant is grown mainly for its delicious seeds used to flavor bread, cakes, candy, and biscuits. The oil obtained from them is similar to olive oil. Sesame will grow in the southern states and should be planted at about the same time as cotton. It is a synergist (increases effectiveness) for pyrethrins.

SEX. The first botanist to demonstrate that flowering plants have sex and that pollen is necessary for fertilization and seed formation was a German, Rudolf Jakob Camerarius, who published his *De Sexu Plantorium Epistula* in 1694. The heated controversy this book engendered lasted a generation before it was finally established that plants had sexual organs and could in fact be elevated to a higher sphere of creation than previously thought.

Bisexual plants. Produce both eggs and sperm. The term also refers to a flower that bears both stamens and pistils.

Dioecious plants. The male and female parts are found on separate plants.

Monoecious plants. Male and female parts are borne in different flowers but on the same plant. A good example of this is the squash plant. Every year readers of my weekly column "Gardening With Louise," published in The Daily Ardmoreite (Ardmore, OK 73401), call me up and tell me their squash plants are blooming but not bearing. The answer is that the male flowers appear first, as the plant grows the female flowers (which will bear the fruits) appear and are fertilized, often by bumblebees. The female flower is borne at the end of a small bulb which gradually enlarges.

SHADE. Two delightful plants for shade that team well with each other are native bleeding heart (*Dicentra eximia*), with lacy foliage and pink, nodding, heart-shaped flowers; and lily of the valley (*Convallaria majalis*), a dainty fragrant perennial with tiny white flowers. A third plant for shady spots is hosta (*funkia*), a plant much valued for its large, decorative leaves.

Achimenes, which come in various colors, is an excellent pot plant for shady locations on the north. It blossoms almost continuously from late spring until fall frost.

SHAMROCK. There has been much argument over which plant is the true shamrock; some say it is a small clover plant with green leaves consisting of three leaflets, others insist it is the wood sorrel. The flowering shamrock sold by florists blooms best in warm weather and with generous sunlight. Water when necessary and do not repot often. Pot binding tends to encourage blooming.

The shamrock is the national flower of Ireland and also appears with the thistle and the rose on the British coat of arms because these are the national flowers of Ireland, Scotland, and England.

Trefoil (*Oxalis Acetosella*), sometimes called shamrock, is a delicate little wild plant which grows in shady places, often in backyards or along fences. Trefoil means "three leaves," each of which is heart-shaped.

SHARING. Gardeners are among the most generous people in the world, and one of the greatest pleasures of gardeners everywhere is sharing—an experience all the more delightful for both giver and receiver if the gift plant is unusual.

An unusual plant should be hard to find in the rank and file of nursery catalogs or garden stores, something not too well known, yet interesting and effective wherever it is planted. If the plant is easily grown and easily separated, so much the better.

Adonis amurensis, a very early spring perennial, blooming sometimes before the snow has entirely melted, is just such a one. The feathery, much-divided foliage dies to the ground by late June and remains forgotten after that; the following spring it peeps through even before the snowdrops appear. It has bright yellow, buttercup-like flowers. The long, fibrous roots are easily divided with a spade. Cut it in such a way that several buds are left on the clump of roots to be used in starting the new plant.

Another gift possibility is the double-flowered bloodroot, a native American plant not, as yet, too well known. The large, white, double flowers appear in early spring not long after the adonis has bloomed. The plant grows well in shade.

Other plants include *Narcissus asturiensis*, a miniature daffodil which flowers in the very early spring along with the snowdrops. A bulb which flowers in early summer is the golden garlic, *Allium moly*. It has yellow, starlike flowers in clusters up to three inches across. White-flowered scillas and grape hyacinths or truly white violets are also good conversation pieces, and these, too, increase rapidly, making them ideal for sharing.

SHOCK ABSORBERS. Many garages will give away discarded shock absorbers for the asking. Place them at the end of garden rows. When watering, slip the hose around them to prevent it from whipping over plants.

SHOE BAG. Get a plastic, see-through shoe bag and use it to store the small separate items such as gloves, plant tags, Twist-ems, in each pocket. This is a time and temper saver, and ideal for the greenhouse.

SHOOFLY PLANT *(Nicandra physalodes)*. Plant this annual for repelling white flies.

SHORT SEASON AREAS. Extend the growing season a couple of weeks or more by using hot caps or plastic row covers for flowers or vegetables that do not grow above eighteen inches high. Also study your seed catalogs and choose some of the many quick-maturing varieties.

SIBERIAN GINSENG *(Eleutherococcus senticosis maxim)*. The plant is so called because *senticosis* means prickly in Latin. One day, it is said, a young Russian doctor named Gorovoy noticed deer greedily eating the leaves of a thorny plant commonly found growing wild in the Russian Far East. It belonged to the Araliaceae and shortly joined the other araliaceous medicinal plants on trial in the laboratory, and it outshone them all, rivaling or even surpassing ginseng itself. It was found to be a first-class tonic plant medicine greatly increasing stamina in long-term administration.

Although similar in pharmacological effect, *Eleutherococcus* was found to have advantages over ginseng. Under the stress of hard physical work both men and animals were found to receive greater stimulation for longer periods with *Eleutherococcus* than with ginseng. In hyperactive individuals it also possesses some calmative effect. Actually the name Siberian ginseng is a misnomer for it is an entirely different plant, though having much the same effect. Its qualities were not discovered until 1962 and there is no mention of it as a Russian folk medicine. It is an example of that rare species, a new national plant medicine discovered by modern research. Since the plant grows abundantly, it is far less expensive to buy than the rarer ginseng.

SIPPING STRAW STILTS. When your cut flowers have stems too short for a tall vase, insert the stems in plastic drinking straws, then cut them to the length you want.

SMOKE. Control aphids, ants, and mites in the greenhouse with smoke from oak leaves which are not poisonous, do not kill soil bacteria, and leave no harmful residue. Dried stems and leaves from canna plants and peppergrass may also be used.

SMOKING. Smoking is not only dangerous to your health, it's dangerous to your plants' health as well. The tobacco mosaic virus also affects members of the tomato family and can be communicated from the hands of a smoker working in the garden.

SNACK-SAC. At last someone has thought about the apartment-house kitty. These delightful kits for cats contain one gallon enriched "peatlite" growing mix and seeds of either catnip or grass for cats (oats). The container is made of heavy duty four mil polyethylene and converts into a charming pot. Oats fulfill the plant-nibbling cat's need for greenery, and provide a safe, healthy distraction from chewing on houseplants (which may in some instances be poisonous). Catnip, a perennial mint, is a feline favorite. The terracotta colored pots are decorated with designs of cats, of course! (Applewood Seed Co.)

SNAILS AND SLUGS ON PERENNIALS. Grayish, wormlike, legless bodies, ½ to 4 inches long when fully grown, hide in damp, protected places during the day. At night they chew up leaves, leaving a glistening trail of slime. Inverted cabbage leaves make good traps for both slugs and snails; or spread wood ashes on the ground. Toads, which eat them, are good garden aids.

SNAKES. The harmless, so-called "garden" variety of snake catches and kills many injurious insects and is, itself, not poisonous. Protect it. In the Southwest be very careful around berry patches, which are attractive to rattlesnakes. Tarantulas, the big hairy spiders, are also attracted by berries.

SNAPDRAGON (*Antirrhinum*). The Greeks named this one *anti*, like, and *rhinos*, snout, to describe the curious shape of the flowers. However, nurserymen are now producing penstemon and azalea flowered types of very different shapes. All are perennials. Two little-known trailing varieties, *A. asarina* (yellow) and *A. glutinosum* (cream and yellow), are delightful additions to the rock garden. Old-fashioned snapdragons are compatible with nicotiana, baby blue eyes, or alyssum.

Snakes can be destructive, but most of them are friends of the gardener, feeding on abundant insect populations.

SNOWDROPS *(Galanthus)*. Winter jewels, undaunted by snow, snowdrops last a long time in bloom. They are lovely in patches of woodland under deciduous trees. Plant in fall so they have a long growing period. Increase your stock right after they bloom; replanting them pays off in a much bigger crop.

SOAP PLANT *(Chlorogalum pomeridianum)* Soap Root.   The powdered bulbs of this native from California and Oregon, are toxic to armyworms and melonworms.

SOAPTREE YUCCA *(T. elata)* Adam's Needle, Candelabra de Dios (Candles of the Lord).   This treelike yucca is often branched and has the tallest flower stalk of any of the yuccas. It is very ornamental with creamy white, lilylike flowers.

There are many species of yucca. The Indians beat the roots of yucca in water, using the milky liquid for a shampoo. Many attribute the Indians' dark hair, well into old age, to the use of this preparation.

SOUR SOP *(Anona muricata)*.   These powdered seeds are toxic to armyworms and pea aphids.

SOWBREAD *(Cyclamens)*.   These are a genus of dwarf tuberous plants from the Mediterranean regions. They are exotically handsome and best grown in a cool temperature. Cyclamen is from the Greek *kyklos*, a circle, referring to the coiling of the flower stems in some species after flowering. This is the plant's method of bringing the seed capsules down to soil level.

Do not keep these plants near any orchid plant at any time as they give off ethylene gases which kill the plant and its blossoms. However, cyclamen has long been esteemed for killing parasites on fruit trees. The principle, saponin, found in the bulbs, is effective fresh or dried.

SOWBUGS or PILLBUGS.   These have oval, dark gray, flattened bodies, seven pairs of legs, and are up to ½ inch in length. They hide under logs, boards, crop refuse, and in other damp places. When disturbed they roll up and look like pills. They feed—and how they feed!—on tender parts of plants and newly emerged seedlings. Look for and eliminate their hiding places. Toads, including the horned toad (really a lizard), are helpful.

SPIDER MITES.   The U.S.D.A. Yearbook *Insects* reports that an emulsion spray containing a 2 percent addition of oil of coriander kills the spider mite. Beatrice Trum Hunter in her book *Gardening Without Poisons* tells us that a 2 percent emulsion spray of oil of lemon grass *(Lippia triphylla)* and *Dracoephalum moldavica* from the mint family is effective against red spiders and cotton aphids. Spider mites attack many flowers, being particularly troublesome on columbine. Experiments at Purdue University made use of an old-time favorite, a combination of buttermilk and wheat flour to destroy spider mites by immobilizing them on leaf surfaces where they "exploded."

SPIDER PLANT *(Chlorophytum comosum variegatum)*.   This interesting houseplant from South Africa develops young plants at the end of its flower stalks. The spider plant needs good light with or without direct sun, a cool room, and moderate humidity. It grows well in ordinary soil if kept moist. Propagate by removing young plantlets and potting them separately. The plants grow one foot tall with wider spread. The leaves are green with broad white center stripes.

SPIDERS.   Just because the note on the spray can says "kills spiders," it doesn't mean that you should. Spiders make good predators and only two common species, the black widow and the brown recluse, are really to be feared. Spiders are rather wonderful—like the nimble crab spider for instance, named for its ability to scurry both sideways and backwards. The little hunter can turn white, pink, or yellow to blend with vegetation. In a mini-jungle of stalks and stems you may see a green lynx spider snatching up a victim as it trails a dragline behind it. This is a safety thread, anchored at intervals, that most spiders put down as they move about.

Spiders are particularly helpful during a heavy infestation of grasshoppers, killing large numbers by trapping them both in their aerial webs and on the ground.

SPINDLE LEAF TREE (*Euonymus europaeus*) SPINDLE TREE.   This deciduous shrub or small tree is wild in parts of Europe. The leaves color well in the fall and its red fruits with orange seeds are very attractive in autumn. The fruit has a paralyzing action on aphids. The wood of the tree was once popular for butcher's skewers.

SPRAY DAMAGE.   When spraying plants with fungicides to control disease, insecticides to control pests, and fertilizers for foliar (leaf) feeding, there is always a danger of damaging the leaves with the spray fluid. This may occur if the spray is not sufficiently diluted, if the plants are in a soft condition, or if the weather and atmosphere are unfavorable. Also, plants growing in a smoky atmosphere may be damaged owing to the liberation of copper by the acids in the atmosphere.

Whenever possible choose a still day for spraying, so that the spray material may be directed where it is needed. This prevents the spray from drifting in the wind and endangering other plants. Do not use oil sprays or other types of spray on fruit trees at blossoming time. The spray might injure bees, bumblebees, or other insects pollinating the fruit.

SPRAYS.   Onions. Red spiders and various aphids, especially those attacking roses, are routed out by onion spray. Grind up onions in a food chopper or an electric blender, add an equal amount of water, strain mixture, and use as a spray. Bury the mash in the flower bed or garden.

Hot peppers. Chop up hot pepper pods. Mix with an equal amount of water and a little soap powder to make the materials stick. This makes an effective spray against ants, cabbageworms, spiders, caterpillars, and tomato worms. Dry hot pepper which has been ground up and dusted on tomato plants, offers protection against many insects. Dry cayenne pepper sprinkled over plants wet with dew is good against caterpillars.

Combinations. Use a combination of several materials as an allpurpose spray. For instance, grind together three hot peppers, three large onions, and a whole bulb of garlic (peeled and chopped). Cover the mash with water and allow to stand overnight. Strain the following day and add enough water to make a gallon of spray. Use on roses, azaleas, chrysanthemums, beans, and other crops three times daily if infestation is exceptionally heavy. Repeat after a rain. Bury mash under rose bushes.

Rhubarb. Rhubarb leaves boiled in water and sprinkled on the soil before sowing wallflowers and other seeds, is a preventive against clubroot. It is also useful against greenfly and black spot on roses.

Tomato leaves. These have insecticidal value as they contain an alkaloid similar to digitalin and more active than nicotine. An alcoholic extract of this substance is very effective against aphids on roses, pears, beans, and other plants. A spray of macerated tomato leaves soaked in water also frees rose bushes of aphids and eggplant of caterpillars.

Elderberry leaves. An infusion made by soaking elderberry leaves in warm water may be sprinkled over roses and other flowers for blight and also to control caterpillar damage.

Soap. A simple soap spray is often effective against aphids, thrips, mites, and other garden pests. Mix one to two teaspoons of Ivory Liquid, Shaklee's Basic H, Tide, or a small chunk of Fels Naptha with a gallon of warm water and apply with a plastic squeeze bottle. Be careful of beneficial insects such as ladybugs and mantises, since the soap may harm them also.

SQUIRRELS. If squirrels dig up your flower bulbs, lay a section of chicken wire on the soil surface of the planted area. Secure with rocks around the edges. This also discourages cats which like to dig in newly cultivated soil.

Protect trees from squirrels with a guard around the trunk. Encircle the tree with a slippery, smooth metal, shaped in the form of a downward cone. Be sure it is wide enough to prevent squirrels from jumping over it.

STAPELIA (*Asclepiadaceae*). The enormous, hairy, star-shaped blossoms of this plant have a real stench. Actually "stink bomb" seems more appropriate than "blossom," for when *Stapelia gigantea* opens one of its blooms, from eleven to sixteen inches across, there is no doubt why it is called the carrion flower. But if you can overlook the odor, the bloom is remarkable—the sort of thing you'd expect to find in a science fiction garden.

There are about ninety species of stapelia, mostly from South Africa, and they belong to the milkweed family. Other, more refined members include hoya and stephanotis, which have sweet smelling flowers, and the ceropegia or rosary vine.

Giant stapelia isn't a good choice for a houseplant, but *Stapelia variegata* is fun to grow and the odor of the strange blossoms isn't nearly as potent. These plants are very attractive, however, to flies.

STAR-OF-BETHLEHEM *(Ornithogalum)*. This early spring bloomer with white blossoms lightly striped with green, is excellent alone, in masses, or as an edging for a bed of daffodils.

STATICE *(Limonium, Goniolimon)* SEA LAVENDER. The stiff flower stems bear literally hundreds of dainty flowers in many-branched panicles. Statice is particularly good for drying as well as being graceful in the garden combined with larger flowers. The plants are superb for seaside gardens as they are unaffected by salt wind or salty soil. They grow well in garden soil; the richer the soil, the larger the flower heads.

STEPPING STONES. If you are planning a flagstone walk, plant creeping thyme *(Thymus serpyllum)* between the stones in a sunny position. The plant seems to thrive from being walked upon and gives off a pleasing fragrance with each step you take.

STINGING NETTLE *(Urtica diocia)*. This cursed weed causes much discomfort when touched. Relieve the itching with yellow dock leaves (usually found growing nearby). Crush and apply as a poultice. Nettle is not all bad, however. In England a pleasant drink called nettle beer is relished by ailing aged folk. And nettle, which is slightly laxative, is a healthy and easily digested vegetable. Boiling renders the stinging hairs harmless, but pick the young tender shoots wearing gloves.

STINKING WILLIE *(S. jacobaea)*. This weed, poisonous to cattle, causes a hardening of the liver.

STRAWBERRY *(Rosaceae fragaria)*. This member of the rose family has small fragrant white flowers followed by delicious fruit. In contrast to the fig, which has its flowers and seeds inside, the strawberry fruit forms its dry yellow seeds on the outside. However, the strawberry (except for some alpine types) does not usually reproduce itself by seeds. During the season when the fruit is developing the plant sends out slender stems called runners which grow on the ground and send roots into the soil.

Onions are as protective to strawberries as they are to other members of the rose family. Lettuce is a good border to the strawberry bed, and pyrethrum, planted near, serves as a pest preventative. A spruce hedge is also protective. Strawberries do well with bush beans, spinach, and borage. White hellebore controls sawfly, and marigolds help if there are nematodes in the soil. Pine needles make a fragrant mulch and also improve the taste of the berries. Spruce needles are also good to use.

STREET NOISES. Trees and hedges muffle street noises. The best trees and shrubs for this purpose are dense evergreens, which also give year-round privacy. Hemlock, yew, and arborvitae are good if they will grow in your area. Keep the lower branches from dying off by pruning carefully; most noise comes from near the ground. If you have a choice of sites when you build, choose one above street level to keep noise at a minimum.

STREWING HERBS. See HERBS, STREWING.

STUDENTS. A recent survey has shown that the most popular plants for students are Swedish ivy, coleus, and spider plants.

STUMPED BY A STUMP? You can make it into a thing of beauty if not a joy forever. Sometimes an old tree must be cut down because it is rotten, yet the stump may not be dug up and removed. Scoop out the middle of the stump and fill it with good soil. In the fall plant tulip bulbs. After they have bloomed in the spring, succeed with petunias, snapdragons, or some other colorful flower.

SUGAR. Sugar kills nematodes by drying them up. Five pounds of sugar per 100 pounds of soil will kill nematodes within twenty-four hours. Helpful fungi are the enemies of nematodes, and plenty of humus in the soil promotes their growth.

SUMAC (*Rhus glabra, Rhus typhina*) LEMONADE TREE. The parts used are the flower heads, picked early to mid-summer, and the fruits as they begin to turn a bright red. Dry for future use. To prepare a tasty summer drink, steep a heaping teaspoonful of the ground flowers and/or fruits, fresh or dried, in a cup of hot water. Cover five or six minutes with a saucer. Stir and strain. The Indians of upper North America sweetened this drink with maple syrup, the tart fruits being soaked in water until needed for use. The fruits of the staghorn sumac are distinguished from those of the smooth variety by being far more hairy; use less of the staghorn because they are more acid.

Bury bags of sumac leaves around the base of apple trees infested with wooly aphids. Tannin has been discovered as an active principle in sumac leaves.

SUMMER FLOWERING SHRUBS. If you want spectacular bloom in July and August, here are some suggestions: hydrangeas (many varieties), *Buddleia davidaii, Ceanothus americanus, Holodiscus discolor, Hyperioum densiflorum, Indigofera amblyantha, Itea virginica, Sorbaria arborea, Stewartia pentagyna, Tamaris pentandra, Abelia grandiflora, Clothra ainfolia, Hibiscus syriacus, Lespedizia bicolor, Perowskia striplexfolia,* and *Vitex negundo incisa.* These do well in the hot and dry areas of the southwestern states.

SUMMER FORGET-ME-NOT *(Anchusa)*.   These plants have blue blossoms and are a fine bedding or filler plant. Use with marigolds and petunias.

SUNFLOWER *(Helianthus annus)*.   This is one of our most valuable flowers. Its bright and cheerful blossoms dutifully follow the sun. The seeds and oil are becoming ever more popular. In the *Medicinal Value of Natural Foods* Dr. W.H. Graves says sunflower seeds "will sustain life indefinitely and are helpful for weak eyes, poor finger nails, tooth decay, arthritis, dryness of skin and hair, and rickets. The oil is soothing to the skin and a good hair dressing in addition to having all the uses of a fine vegetable oil for cooking and salad dressings. Contains vitamins B_1, A, D and F." What more can you ask of a plant?

Also try sprouting the black sunflower seeds and using the sprouts with other greens as a salad. The black seeds are more delicious and nutritious than even the striped ones.

SUN-SCREENS.   Too much of the sun's harsh rays dehydrate and dry out your skin and reinforce the lines that come from squinting. Sunglasses protect your eyes from glare and discourage frown lines around the eyes. When working in your garden, use creams and cosmetics that contain generous amounts of sun-screens such as PABA (para-amino-benzoic acid), a B-vitamin derivative that guards against premature wrinkling and skin aging.

SWAN RIVER DAISY *(Brachycome iberidifolia)*.   This native of Australia is pretty in rock gardens with poppies and sedum, or used as an edging plant. Plant in masses. The name comes from the Greek, *brachys*, short, and *comus*, hair. Named hybrids include Little Blue Star, "Purple Splendor," and "Red Star."

SWEET FLAG *(Acorus calamus)* CALAMUS.   The alkaloid root works as a contact poison to insects, even though it is edible to humans. It commonly grows in swamps and along brooks.

SWIMMING POOLS.   You can have a carnival of color without the pain of constantly cleaning up, if you choose wisely for your poolside planting. And don't plant anything bristly, sharp, or thorny. Plants should also be as litter-free as possible.

Shrubs. Here are some beautiful but practical suggestions: Camelia, *Crassula argentea, Fatsia japonica,* Griselinia, Juniperus, Raphiolepis, *Viburnum davidii.*

Trees. Cordyline, Dracaena, *Ficus auriculata, Ficus lyrata, Firmiana simplex,* Musa, palms, Schefflera, Strelitzia, tree ferns.

Vines. *Beaumontia grandiflora,* Cissus, *Fatshedera lizei, Solandra maxima,* Tetrastigma.

Perennials. Agapanthus, *Agave attenuata, Aloe saponaria, Alpina zerumbet,* artichoke, *Aspisdistra elatior,* canna, *Clivia miniata,* cyperus, Gazania, Hemerocallis, *Kniphofia uvaria,* Liriope, Ophiopogon, philodendron (treelike types), Phormium, sedum, succulents, yucca, *Zoysia tenuifolia.*

Remember that the litter produced by plants (and this should be reduced as much as possible), should be large enough to be removed by hand rather than by passing into the pool's filter.

SYNERGISTS. Usually derived from plant products, synergists are non-toxic substances which are added for a strengthening effect. Pyrethrum, for instance, is greatly strengthened by the addition of asarinin from bark of southern prickly ash, sesamin from sesame oil, and peperine from black pepper.

T

TANSY *(Tanacetum vulgare)* PARSLEY FERN. Tansy has a very strong, bitter aroma. The Latin *tanacetum* derives from a Greek word indicating immortality because the dry blossoms do not wilt. The distilled oil re-

Tansy is a friend of both vegetable and flower gardeners, since it will repel ants, borers, cucumber beetles, Japanese beetles, and squash bugs.

pels flies and mosquitos. The plant has been used against intestinal worms (oxyuri) and in wine against stomach and intestinal spasms. The Russians used it as a substitute for hops in beer and rubbed it on the surface of raw meat to protect it from flies. Tansy planted near the entrance to the house deters ants.

TELEGRAPH PLANT *(Desmodium gyrans).* This native of India belongs to the pea family, Leguminosae. It has trifoliate leaves; the center leaf is elliptical and about two inches long, the two side leaves are about ½ inch in length. The leaves are usually in constant motion, rising and falling alternately, but not in regular time. The rise and fall of the leaves has been compared to railway telegraph signals. The plants are most active in early morning, especially the young ones. They make a wonderful conversation piece.

Use sandy soil when sowing the seeds of annuals in pots or when rooting cuttings.

TERMINAL SHOOT. This term indicates the shoot that forms the end of the main stem or of a main branch of a tree or plant. All other shoots are called side shoots or *laterals.*

TERRARIUMS. Terrariums fascinate and are not at all difficult to care for. Mix up soil for the planting base. A good combination for foliage plants and a general woodsy scene is two parts peat moss, one part perlite, two parts reasonably decent garden soil (or potting soil), and one part sand. You can buy a ready-mix from a nursery and adjust according to the kinds of plants you will be growing.

Clean your container thoroughly. Scrub, rinse, and air if it has been used before. Put down the bottom layer of an inch or so of moist perlite (the exact depth depends on the proportions of the container). Sprinkle a thin layer of charcoal on top of the perlite. Now add the soil mixture and do whatever you want for basic land construction such as hills, valleys, or gentle slopes.

Tentatively place the plants in the terrarium to make sure that the living design is effective. Then plant your miniature garden. In large containers you can sink the pots to their rims. In smaller terrariums either unpot the new plants and shake their old soil loose, or set them in the new soil and tamp it around to the same level as the pot soil. Or tap the plant loose from its pot and plant the entire earth ball. The choice depends on the quality of the original pot soil and the individual plant's sensitivity to transplanting.

Plant ferns and small evergreen seedlings for the taller growth of the terrarium. Ivy, moss, and lichens are also attractive. For a grasslike, "carpet" effect, plant partridge berry. Place each plant carefully in the soil with enough space between plants to allow for growth. Set the completed terrarium in a light place, but not where sun will strike it. With the glass container closed, this "balanced terrarium" preserves temperature and moisture inside. Open the lid if the glass clouds with moisture.

TETRAPLOIDS. Seeds or plants exposed to atomic radiation are often grossly changed. A drug, *colchicine*, extracted from the fall-blooming crocus, profoundly changes plants too, sometimes doubling the number of chromosomes in every cell so that a normal diploid becomes a tetraploid. Such plants have enormous vigor, larger blossoms, or bigger fruit.

The brown-eyed Susan, a meadow plant, was treated with colchicine and changed into the vigorous and beautiful gloriosa daisy, a tetraploid. Other examples are blue mink ageratum, glamour phlox, and tetraploid snapdragons and zinnias. The small French marigold has double the chromosomes of the large American type. The result of crossing is an intermediate type, a triploid that is sterile; because it fails to produce seed it continues blooming with tremendous vigor all season.

TEXAS BLUEBONNET *(L. texensis)*. The state flower of Texas is beautiful grown in masses. It belongs to the Lupine family, and the spikes of sweet-pea-like flowers are a bright, rich blue. Nitrogen-fixing legumes, they thrive on well-drained, poor, sandy soils in full sun. These plants are easy to raise from seed but difficult to transplant.

THERAPY. Many professional persons are advocating gardening to alleviate mental depression. They believe that gardening is a time-proven way to stay "alive and well," both mentally and physically. If you know someone who is lonely or depressed you might write on their behalf for information from National Council for Therapy and Rehabilitation Through Horticulture, Sunland Center, P.O. Box 852, Marianna, Florida 32446.

THISTLES *(Compositae)*. Thistles, though beloved of butterflies, have never been popular with people. In spite of their beautiful flowers, the prickly leaves are unappealing. Thistles are rich in potassium (good in the compost heap), and would have high feeding value if it were not for their thorns. In grain fields they take away food and moisture, and in pastures they protect and thereby increase the spread of other weeds.

To get rid of thistles, timing is important. If cut before the blossoms are open the thistles will spread from the rootstocks. If cut after the blossoms are pollinated the situation is a little better. But, if shortly after pollination, the blossom-heads only are cut off, the plant will bleed to death and wilt.

Thistles, beautiful when in bloom, are bothersome in grain fields and pastures. Compost them. They're rich in minerals.

THUNDERGOD VINE *(Tripterygium wilfordii)*. This plant is a great favorite in China as an insecticide. For centuries Chinese gardeners have sprinkled their gardens with a powder made from its roots to keep insects away.

THYME *(Thymus serpyllum* and *T. vulgaris)*. This very valuable plant has been used in medicine since the very earliest days of herbal treatment. It is a powerful antiseptic and general tonic. As an aphrodisiac thyme crops up with almost monotonous regularity in literature throughout the ages.

Thyme yields an essential oil which accounts for its antiseptic properties and which is a good vermifuge. The oil, called *thymol,* is found in many orthodox preparations such as disinfectants, dentrifices, and hair lotions.

TILLANDSIA. These tender, evergreen plants have attractive flowers and large, beautifully colored bracts. They belong to the Bromelia or pineapple family. *T. usneoides,* the Spanish moss, is another family member.

TOAD-FLAX *(Linaria vulgaris).*   Toad-Flax, found in waste places and often growing among corn, has powerful dissolvent properties and is therefore used to treat obstructions in all parts of the body, particularly the intestines, kidneys, and bladder. The leaves are small and flat; the flowers are in racemes of yellow and orange, marked white, and of the familiar snapdragon form. The plant is also considered one of the best jaundice remedies known to the herbalist.

TOADS.   Toads are very helpful in the garden for catching and eating unwanted insects. They work long hours for low pay. If you cannot find any, they can be purchased commercially.

TOBACCO.   As a general pesticide tobacco has long been in use; Olds, Shumway, Stokes, and Henry Field carry seeds if you want to grow your own. Powdered tobacco is recommended for pot plants. The tame nicotiana is not dangerous to your health.

Toads, like snakes, work for the gardener by eating insects. You can introduce toads to your garden, and they'll remain if they have lots to eat and some shelter, such as a piece of a clay pot.

TOMATO HORNWORM. The tomato hornworm is repelled by borage, marigold, and opal basil.

TOMATO PLANT EXTRACT. Boil stems and leaves in water. When liquor cools, strain and spray or sprinkle over plants to destroy black or green flies and caterpillars. The peculiar odor left behind also seems to deter insects from returning.

TOOLS. Drill a ¼-inch hole in each handle and hang the tool over a headless nail driven into the wall. To get a large number of tools in a small area, space the nails close together. Also paint tool handles orange—the color scientists consider safest for hunting caps and jackets. The color is readily discernible to the eye when you lay a trowel or clippers down somewhere in the garden or grass.

TORTOISE. This turtle lives in arid regions. They are large, timid, harmless animals. But they do eat plants and may ruin low-growing tomatoes by taking a bite out here and there. If you find one in your garden, do not kill it; remove it to another area.

TRANQUILIZERS, NATURE'S OWN. Pleasant at any time, herbs are particularly appropriate at times of emotional stress. Herbs that are exceedingly soothing to the nervous system are camomile, valerian, rosemary, and lavender. For an herbal bath put the herb or herbs on a muslin square and tie them up like a hobo pack, toss it into the bath as the water runs in and allow to steep for at least ten minutes. Also good are pine needles, fresh from the tree, wrapped and used like the herbal bath bags. And, while you are relaxing, drink a cup of warm camomile tea, one of the oldest nerve-calming teas beloved of herb enthusiasts— far better for you than tea or coffee. Verbena tea is also highly thought of.

Jethro Kloss (*Back to Eden*) has some additional helpful foods: celery, dill, fit root (scullcap with goldenseal and hops), lobelia, motherwort, origanum, pennyroyal, red clover, rosemary, rue, sage, spearmint, St. John's-wort, thyme, vervain, wild cherry, wood betony, blue violet, sanicle, buchu, red sage, catnip, peppermint, marshmallow root, and mugwort (make an antispasmodic tincture for quick results).

TRANSPLANTING TIPS. Flower and vegetable plants can crowd each other when seeds are planted too closely together. Crowding delays maturity, stunts growth, and distorts the roots of carrots and other root crops. At the same time there may also be skips; transplant enough seedlings to fill the skips, then discard the surplus.

Here's how to transplant so there is as little shock as possible:

1. Start when the plants are small; if they have four to six leaves, they're big enough.
2. Transplant at sundown on a cloudy day. Wind can injure as much as sunlight.
3. Wet soil thoroughly around the roots of seedlings that are to be moved. Do this a few hours beforehand so that the plants will be plump with water.
4. Dig transplanting holes before you uproot any seedlings. Fill the holes with water and let it soak in.
5. Shove a trowel in deeply to pry up plants. Move seedlings with as much soil around the roots as possible.
6. Move one plant at a time. Transplant quickly; don't delay.
7. Immediately soak the soil around each transplant. Don't wait until you have completed the row. For a week thereafter, sprinkle the transplants at least daily.
8. Never apply garden or houseplant fertilizer, liquid or dry, around newly transplanted seedlings. Their roots are too damaged to take it up.
9. If you must move large seedlings with many leaves, trim back half the foliage to reduce the leaf area through which water is lost. Use a shovel to move a big rootball and try to keep it from breaking up when you set it in place.

From the Troy-Bilt Owner News comes an excellent transplanting tip by Carl N. Anderson of Windham, NH. "When raising and transplanting seedlings in the house or greenhouse I've found an ordinary table fork is the ideal transplanting tool. It will loosen the plants in the seed flat without damaging the roots; it will open a hole for the new transplant in the new flat or pot by rocking it sideways; and then by sliding the tines around the delicate stem and pressing down, the transplant is firmed in the growing medium. I have a very high survival rate using this method."

TRANSPORTING PLANTS. Carry cut flowers such as iris in containers made from beverage cans. Punch six or seven holes in the top with a can punch, partially fill with water, and insert stems. The can holds the flowers upright.

TRAP CROPPING. Some plants earn their living by repelling potential trouble, others by luring insect pests away from more valuable plants. Alfalfa lures lygus bugs away from cotton plants; dill attracts the tomato worm; Japanese beetles are attacted to white or pastel zinnias, white roses, or odorless marigolds; mustard attracts the harlequin bug; nasturtiums attract aphids; eggplant attracts the potato bug; radish plants attract flea beetles; and radish with squash attracts the squash bug.

TRAVELERS' TIPS. To traveling gardeners: resist the temptation to dig up unusual plants. Certain plants may not be shipped to certain states and there is sound reason for this. Whenever a species, plant, animal, or insect, not native to the environment and where natural controls are not present, is introduced, Big Trouble may lie ahead. Given the right conditions, it can spread like wildfire. Just look at the case of the Mediterranean fruit fly, the present worry over fire ants in certain southern states, and the so-called "killer bees." The beautiful water hyacinth, now clogging southern waterways, is another case in point. By bringing plants home and possibly letting them escape from your garden you may be creating a monster.

TREES. Southern gardeners delight in magnolias, but farther north the autumnal foliage of such trees as silver, red, and sugar maples; red and white oaks; and white birch brings a similar pleasure.

For flowering trees there are black and honey locusts as well as a new one, bristly locust, which has large deep rose-colored flowers from late May to mid-June. Other "blossomers" include eastern red bud, tulip poplar, wild black cherry, catalpa, white flowering dogwood, and little leaf linden with its inconspicuous but very fragrant flowers. (Musser Forests)

TREE TANGLEFOOT. Tree tanglefoot is a non-drying sticky compound which forms a barrier against climbing and crawling insects such as caterpillars and cutworms. It is effective used on trees.

TRUMPET-CREEPER (*Campsis radicans*). This woody, high climbing vine is very hardy. Although a native of the woods, it is often planted in gardens. The flowers are three-inch-long, orange tubes with flaring scarlet lobes, and grow in clusters. They yield copious nectar and attract many insects. The shade of red orange is somewhat harsh so use this vine with discretion. The fruit is a long pod with a many-winged seed. Some people are allergic to the leaves and may get dermatitis from touching them.

TULIP (*Tulipa*). These beautiful bulbs suppress the growth of wheat. In Holland, the Tulip Festival, which takes place in May when the flowers bloom, is a renowned event.

TULIP TREE (*Liriodendron tulipifera*). This near relative of the magnolia has showy yellow blossoms that resemble garden tulips in size and form. A good shade tree with fast growth, it is ideal for young folks who have just purchased their first home or for older people when they move

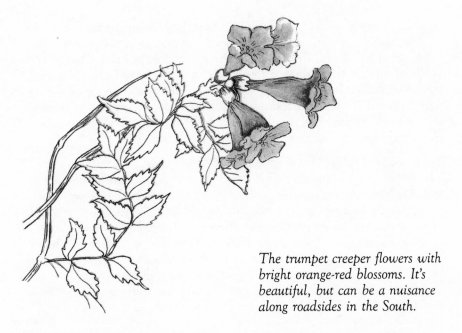

The trumpet creeper flowers with bright orange-red blossoms. It's beautiful, but can be a nuisance along roadsides in the South.

to treeless suburbia. In the lumber trade tulip tree is called yellow poplar.

TURKEY-MULLEIN *(Eremocarpus setigerus)* DOVE WEED. Greenish flowers; dark gray, shining seeds; and stinging hairs characterize the turkey-mullein. Since the leaves contain a narcotic poison, the Indians used the foliage to stupefy fish and poison their arrow points.

Other plants used to stun fish were: blue curls *(Trichostema spp.)*, vinegar weed or camphor weed, wild cucumber *(Marah spp.)*, and members of the gourd family. To do this, pieces of green root were placed in streams. Turkey-Mullein is also toxic to cross-striped cabbage worms.

U

U-GARDENS. Certain flowers, called sensitive flowers, open in the morning and close in the evening. People once thought that the sun's rising and setting was the reason that these flowers opened and closed each day. But scientists have discovered that these flowers have a natural "clock" inside them.

During the 1800s, people planted sensitive flowers in U-shaped gardens. These gardens were called flower clocks because people told time by them.

One point of the U was planted with the spotted cat's ear flower, which opens at 6 in the morning. Then five more different flowers were planted in the row, each opening one hour later than the flower before it. The passion flower, opening at noon, was placed at the center of the curve of the U. From this curve to the other point of the U, six other kinds of flowers were planted in a row. Each one *closed* one hour later than the flower before it. The last flower was the evening primrose, which *opened* at 6 in the evening. The flower clocks were not as exact as the clocks of today, but the inside clocks of the sensitive flowers made them open or close at nearly the same hour every day.

UNICORNPLANT *(Proboscidae)* DEVIL'S-CLAWS, ELEPHANT-TUSKS. The showy, reddish purple to coppery yellow flowers are large and attractive but few in number. More spectacular are the large, black, woody pods ending in two curved, pronglike appendages that hook about the fetlocks of burros and the fleece of sheep. In this way, the pod is carried away from the mother plant and the seed is scattered. The attractive pods are used for many decorative purposes; some even are painted to resemble birds. Young pods are eaten by desert Indians as a vegetable. The mature fruits are gathered by the Pima and Papago Indians, who strip off the black outer covering and use it for weaving designs into basketry.

UVA URSI *(Arctostaphylos uva ursi)* BEARBERRY, MANZANITA, WILD CRANBERRY, BEAR'S GRAPE, SAGCKHOMI. This low-growing evergreen has pretty pink flowers from April to June. The leaves are very useful in diabetes, Bright's disease, and all kidney troubles. A tea is made by steeping a heaping teaspoonful of the dried leaves in a pint of boiling water for thirty minutes. Drink ½ cupful every four hours.

Plant in early fall or spring in land that is loamy and free of lime.

V

VANILLA *(Vanilla).* The extract from this group of climbing orchids is used to flavor chocolate, ice cream, pastry, and candy. The vanilla vine, cultivated in Mexico for hundreds of years, has been introduced into other tropical areas, mainly Madagascar, and the Comor and Reunion Islands. However, it is said to set seed *naturally* only in Mexico. Elsewhere it must be hand pollinated, adding greatly to the cost of its production. The cultivated plant lives for about ten years, producing its first crop at the end of three years.

The flowers, though dull in color, are very fragrant. Vanilla is obtained from the prepared seed capsules of *V. fragrans (planifolia)*, which are six inches long and beanlike in shape. To grow vanilla in a greenhouse, a tropical atmosphere is required; in winter a temperature of 60° F. is suitable.

VEGETABLE SPAGHETTI. This squash variety is an excellent low-calorie substitute for spaghetti. It's easy to grow and easy to cook. Bake or boil for 30 to 45 minutes, then scoop out.

	Regular spaghetti	Vegetable spaghetti
Calories	155	95
Carbohydrates	32	23
Calcium	11 mg.	49 mg.
Vitamin A	0	12,690 I.U.
Vitamin C	0	14 mg.

VERBENA *(Verbenaceae).* Vervain or verbena was the holy herb used in ancient secret rites; it was also supposed to cure scrofula and the bite of rabid animals, to arrest the diffusion of poison, to avert antipathies, and to be a pledge of mutual good faith—hence it was worn as a badge by heralds and ambassadors in ancient times.

Most of our perennials come from South America and are hardy only in favorable climates. They come in many colors and types, and are nice for edging or hanging baskets. Verbena is attractive grown with yarrow and dusty miller.

VICTORIAN BOUQUETS. The Victorians were lavish with flowers. They planted large formal bedding gardens for viewing and for cutting, and grew quantities of flowers which they arranged in huge bouquets of many varieties. Many gardeners are now copying the Victorians by mix-

A wave of nostalgia is sweeping the country, and this is making Victorian gardens very popular. Victorian flower arrangements, too, are seen more often.

ing many plants in one planting container, for a full summer of floral abundance.

Choose flowers of different shapes, sizes, and forms. Arrange so that tall plants are surrounded by less stately plants. Fill out with varieties that will tumble over the side of the hanging basket or planting container. Flowers growing like this create their own sense of compatability and usually grow well together.

Choose a container with ample volume for supporting the growth of ten or twelve plants of four or five varieties at a minimum. To keep the plants growing well, give them plenty of water and fertilizer, or use soluble fertilizer as you water.

Plants to include in your Victorian bouquet might be: pansies, browalia, *Felecia ameiloides*, verbena, lobelia, dianthus, zinnias, begonias, ageratum, marigolds, vinca or snapdragon, torenia, petunia, impatiens, coleus, and cinerarias. For height choose from such winning plants as snapdragons, geraniums, Shasta daisies, or the taller varieties of African marigolds.

VICTORIAN GARDENS.   The Victorian style garden is again becoming popular. This style provides for a garden of seclusion and natural beauty. Such gardens include lawn shrubbery (including the exotics), terrace walks, and even a conservatory if room permits.

VICTORIA REGIA (*Nymphaeaceae*).   The plant was named Victoria regia in honor of Queen Victoria. The round leaves with upturned edges measure up to seven feet across and are strengthened by a marvelous network of veins capable of sustaining weight up to 150 pounds. A child can easily sit on the leaves.

The huge flowers are nocturnal; it is a breathtaking sight to watch them open in early evening, rapidly moving from a bud to a creamy white, wide open, deliciously-scented flower. Closing the next day at about noon, they open again three or four hours before dusk, with the color turning to a definite pink. They fade the next morning and sink below the surface of the water.

VINES.   There is always a place in every flower garden for a truly outstanding vine. Then, too, vines are often used to screen out an unsightly area, to create shade, or as protection for other plants. Many vines add grace to hanging baskets.

Some suggestions for annuals are: cathedral bells (cobaea scandens) thirty feet; Cypress vine (*Quamoclit pennata*), various colors, twenty-five feet; cardinal climber (*Ipomoea quamoclit sloteri*), scarlet, white throated tubular flowers, thirty feet; marble vine (*Diplocyclos palmatua*), attractive leaves; black-eyed Susan vine (Thunbergia); and white wings.

*Virginia creeper is often mistaken
for poison ivy. Note the five
leaflets on the Virginia creeper.
Poison ivy has three.*

Perennial vines of interest are: coral vine (*Antigonon leptopus*) 30 feet;
bignonia radicans (Campsis), red trumpets; Chilean jasmine (*Mandevil-
lea suaveolens*), fragrant white flowers; butterfly pea (*Clitoria ternatea*),
double light blue flowers; Dutchman's pipe (*Aristolochia elegans*) (calico
flower), remarkable pipe-shaped flower; flag of Spain (*Quamoclit lobata*),
crimson flowers, heart-shaped leaves; perennial sweet pea (*Lathyrus lati-
folius*), many colors; Madeira vine (*Boussingaultia basselloi*), vigorous
climber, fragrant white flowers; wisteria, blue or purple clusters in
spring; queen of vines; and yellow jasmine (humile), a delightful evergreen bearing many clusters of bright yellow bells in summer.

VIOLA (*Violaceae*).   Viola is one of the really tough plants; use as pot
plant or in borders or for bedding.

VIOLETS (*Violaceae*).   Nelson Coon in his work, *The Complete Book of
Violets*, says some very interesting things about violets. For instance,
"There are stemmed and stemless violets. Stemmed violets have leaves
and flowers growing from a central stem. The stemless violets' leaves and
flowers grow directly from the tip of the rootstalk.

"Violets often bear two types of flowers, open ones in the spring, and
later, closed or cleistogamous ones, without petals. The open flowers

Violets are a favorite flower of almost everyone. They're beautiful, no mat-ter which species is grown, and the blossoms are rich in vitamin C and can be eaten.

have five petals, two upper, two side, and one lower, usually flat with guidelines for bees. These petals are of various colors, yellow, white, and shades of lavender and purple, sometimes with two contrasting colors on the same flower. The closed flowers produce self-pollinated seeds if cross-pollenization has not occurred."

Violets are among the "artillery flowers"—the seed pods, when ripe, split apart and the seeds are flung hither and yon to begin new plants.

Violets with their delightful, fresh, springlike fragrance are so rich in vitamins C and A that Euell Gibbons (*Stalking the Healthful Herbs*) calls them "nature's vitamin pill." The violet blossoms, which are edible, are three times as rich in vitamin C, weight for weight, as oranges.

Violets are used in many delicious recipes, which include violet syrup, candied violets, and even a violet bombe made with candied violets, ice cream, and whipped cream.

VIPER'S BUGLOSS (*Echium*).   These plants are biennials, usually blue with gray green foliage. They are fine for rock gardens and especially so for seacoast gardens. For a pretty combination, plant with columbine or armeria.

VIRGIN'S BOWER (*Clematis*).   There are a number of colors and vari-eties of these rather small, deciduous vines, highly prized for their often showy flowers. Give virgin's bower a place on fences, trellises, or posts. It likes a cool, shaded root in rich soil and full sun to partial shade.

This is a marvelous companion plant for enlivening the rather somber branches of pine trees; Japanese gardeners often train it to grow like this. It is not a parasite and does not harm the tree.

Clematic jackmani bears four- to six-inch purple flowers; Duchess of Edinburgh, magnificent pure double white; pink chiffon, an exquisite shade of shell pink; ramona, big and blue six- to eight-inch flowers which are among the largest and bluest. All are truly spectacular when in full bloom.

VITAMIN A. See CARROTS.

VITAMIN E. No other vitamin has ever provoked such controversy as vitamin E; some authorities consider it helpful for practically everything, others vigorously deny it. But in case you're interested, here are the richest natural food sources of vitamin E: kidney and navy beans, red beet tops, broccoli, Brussels sprouts, carrots, celery, corn oil, kale, lettuce, oats and oatmeal, olive oil, parsley, peanut oil, soy oil, spinach, turnip tops, wheat germ, whole wheat flour, and watercress.

VITAMIN ROSES. Roses are the flower of love and none more so than the vitamin-packed rosa rugosa. These roses have the large, meaty rose hips—a treasurehouse of life-giving vitamin C. To make rose-hip marmalade, soak the cleaned rose hips for two hours in plain cold water; then let simmer for two hours, and strain. Measure the purée, and add one cup of brown sugar to each cup of rose-hip purée. Boil down to thick consistency. Pour into sterilized glasses and seal.

W

WALLFLOWER (*Cherianthus*). This perennial is usually orange or golden. Before sowing seed, water drillings with a spray of rhubarb leaves boiled in water to protect against clubroot.

Combine wallflowers with daffodils and tulips; in cool climates combine with snapdragons and dusty miller. New strains exist in lovely pastel shades of cream, lemon, apricot, gold, salmon, light pink, rose, ruby, purple, copper, and rust. Wallflowers are good grown with apple trees.

WALNUT, BLACK (*Juglans nigra*). The odor of black walnut leaves is repellent to insects.

WANDERING JEW (*Tradescantia*). The endearing habit of all the wandering Jews is that they are luxuriant in their growth habits. They are many-branched with a compact leafing pattern that is enhanced by a variety of color forms—from shades of green, to variegated white and green, to green and red.

Wandering Jews do best under filtered light because they originate in the rain forests of the tropics and semi-tropics. Use potting mediums of equal parts of loam, peat moss, and perlite. Pinch them back occasionally to keep them looking tidy.

WARTS.   A number of plants are credited with eradicating warts. A Romany remedy is the thick white juice exuded by milkweed plants. Other cures include: the oil from castor bean plants, the white milklike sap of dandelions, or the milky sap of figs.

WASPS.   The parasitic wasp trichogramma is an efficient destroyer of the eggs of many moths and butterflies which are leaf eaters in the larval stages. It is also effective against the eggs of cabbage worms and loopers, corn earworm, and geranium budworm.

Encarsia wasps are tiny and parasitic to white flies. Aphytis wasps are live controls for scale insects. Fig wasps, which live in caprifigs, carry pollen from the male flowers to the female flowers of the Smyrna figs. The calimyrna, a variety of the Smyrna, is widely grown in California.

Black-eyed peas and other field peas grown widely in the southern states, are largely pollinated by wasps. Gather the pods early in the morning while the wasps are still lethargic from the cool of night.

WATER.   We know that seeds and the roots of growing things need water, but the water also carries nourishment. Sometimes elements for plant nourishment are in the rain itself, sometimes plant food previously put in the garden is dissolved by the rain. The most fertile soil

Wasps, often ill-tempered and prone to sting, are very effective against a variety of insect pests. They're helpful as pollinators, too.

and the best climate conditions cannot produce a single green leaf if there is no moisture.

It is important to use water wisely, especially in areas with little rain. To conserve moisture, soak furrows before planting seeds, and then cover with dry soil. Though invisible to the eye the moisture is there where the seeds need it to germinate.

When setting out new nursery stock, especially in dry weather, dig the hole and water deeply several times before planting the shrub or tree. Then the water will be available to the roots. Splashing a little water on top after setting won't do the job. When the hole is half-filled with soil, pack firmly, and fill with water. As water soaks down, fill in with soil and leave a saucerlike depression, putting a layer of loose mulch in the depression.

Mulch is important over all the garden; it prevents a crust from forming and lets water soak down easily and more slowly, preventing runoff. Check soil under mulch to see if watering is needed.

For vine crops such as tomatoes or even decorative vines, it's good to sink a perforated tin can on one side of each plant. Whenever the plants need water, fill the cans. Once you start watering, continue until there is a good rain, or you may lose your plants.

WATER LILY *(Nymphaeaceae)*. Water lilies, as with many other plants, have been hybridized and now come in an almost endless variety of magnificent creations, beautiful in form and color. The lilies mentioned here are less showy but very interesting in their own way.

White pond lily *(Nymphaea alba)* grows wild on ponds, lakes, and other still waters. The name is from the Greek for water nymph. The flowers are large, solitary, rounded of form, and sweetly scented with prominent yellow stamens. The root is soothing and astringent with antiseptic properties. The leaves are sometimes used for binding over wounds or inflamed areas of the skin.

The yellow pond lily *(Nuphar lutea)* is also medicinal. Its common name is "brandy bottle" from the brandylike scent of its flowers and the shape of its seed vessels, which are like the traditional brandy flagons. These lilies grow wild in the shallows of lakes.

WEATHER FORECASTERS. Old sayings or clues from nature often contain truths. Signs of a hard winter include: unusually large crops of nuts or acorns, heavy moss on north side of trees, when sap of maple and sassafras goes down early in the fall, when leaves of grapes turn yellow early in the season, and thick husks on the ears of corn.

When the flowers of scarlet pimpernel close during the day, it was believed to be a sign of rain. For this reason the plant also became known as "poor man's weatherglass."

Mushrooms and toadstools are said to be more numerous just before a rain.

If the down flies off dandelions when there is no wind, it is a sign of rain.

Scientists are studying plants in areas known to be earthquake-prone, believing that certain plants by their behavior indicate when tremors are impending.

WEEDS. Some weeds have been found helpful in flower beds. Lamb's-quarter gives added vigor to zinnias, marigolds, peonies, and pansies. A carpet of low-growing weeds from the despised purslane family growing among rose bushes improves the spongy soil around their roots. Lupine, a legume, helps corn and other cultivated crops. Morning glory planted near corn enhances root vigor. Wild mustard is beneficial to grape vines and fruit trees. Small amounts of yarrow and valerian add vigor to vegetables.

WEIGELA *(Weigela)*. This lovely shrub flowers from May through July. The rosy blossoms resemble foxglove in shape, are borne in immense quantities, and are attractive to hummingbirds. Give them a moist soil and full sun, away from competition of tree roots. As they bloom on twigs of the preceding year, prune after flowering.

WELWITSCHIA. This strange African plant has a short, woody trunk rising from a large taproot and spreading like a table top to a width of five or six feet. The plant resembles a giant flattened mushroom. A single pair of green leaves spills over the top. They are two or three feet wide and often twice as long. They live as long as the plant does but the hot winds split them into long, ribbonlike shreds which trail on the dry ground.

WHEELCHAIR GARDENING. Choose easy to grow plants which need a minimum of repotting. Terrariums, midget gardens, dish gardens, and bonsai are all practical choices for wheelchair gardening.

WHITE FLY. White flies are repelled by nasturtium, marigold, and nicandra (Peruvian ground cherry).

WHITE HELLEBORE *(Veratrum)* FALSE HELLEBORE. This was a safe, popular insecticide against slugs, caterpillars, and other leaf-eating pests in early American gardens. It was used as a dust, or dissolved for a spray: one ounce to three gallons of water.

WHITE MARIGOLDS *Burpee's First Whites.* "These are the world's most famous marigolds," according to the Burpee catalog, "a mixture of white and nearly white types." Since First Whites bloom rather late, the Burpee people recommend sowing the seeds six to eight weeks indoors before outdoor planting time. The plants average twenty-eight inches in height.

WHITE-THORN ACACIA *(Acacia constricta).* This acacia has fragrant masses of yellow flowers, amply protected by long, straight, white thorns. It grows in Texas, Arizona, and Mexico, and makes a good barrier plant for traffic control.

WICK WATERING. When you vacation, provide your larger container plants with wicks leading from the pot to a water supply in a coffee can fitted with a plastic lid. Punch a hole in the lid, then thread the wick through. Lid cuts down on evaporation so water supply lasts longer.

WILD BIRDSEED. Some birds are almost exclusively seed eaters; others are "switch hitters," eating insects as well as seeds. Among the seed eaters are finches, nuthatches, titmice, sparrows, siskins, towhees, juncoes, jays, Clark's nutcracker, and of course, doves, pheasant, and quail.

Depending on your location and altitude, here are some planting suggestions for birdseed: coreopsis, cosmos, sunflowers, verbenas, and thistles as small flowering plants; burnet and croton as dove favorites; trees such as spruce, fir, birch, pines, oaks, and palo-verdes. In fall birds love the hackberries. Native grasses having seeds beloved of birds are: bluegrasses *(Poa* species), gramma grasses *(Boutelous* species), bluestems *(Andropogon* species), wheatgrasses *(Agropyron* species), vine mesquite *(Panicum obtusum),* and Indian ricegrass *(Oryzopsis hymenoides).*

WILD CHERRY TREES. The avid tent caterpillar likes to eat wild cherry trees. Entymologists have discovered that if an insect is deprived of its native feeding plant and learns to eat another plant, it will never return to its original feeding plant. What this means is that if all the wild cherry trees were destroyed, the tent caterpillar would go to other trees, principally apples and pecan. Worse than that—it would never go back to the wild cherry. Keep this in mind and don't destroy the wild cherry trees; they are valuable as trap plants for concentrating the tent caterpillars' enthusiasm for leaves where they can do little harm. Even when they are completely defoliated, the wild cherry instinctively protects itself against permanent damage. In about three weeks it may be again in full leaf.

The enemies of the tent caterpillar, the calosoma beetles and braconid wasps, as well as other insect friends, help keep tent caterpillars under natural control.

WILD CUCUMBER *(Echinocystis fabacea)* MANROOT. The powdered root is toxic to European corn borer larvae.

WILD MUSTARD (*Brassica arvensis*).   Wild mustard is fairly common in cultivated areas and waste places. Gather the young leaves in spring and they may be steamed, or used in a cold salad or soup. Wild mustard growing among fruit trees or grape vines is beneficial, according to Beatrice Trum Hunter in her book *Gardening Without Poisons*.

WINDFLOWER (*Anemone*).   The name is from the Greek *anemos*, wind, and *mone*, habitation. The plant is so called because some species are found in windy places. The windflowers are suitable for the border, for the rock garden, and for cutting. Their blue, pink, and white coloring combines well with narcissus.

WINDOW BOXES.   Here are some popular plants suitable for many types of window boxes. In sun try geraniums, lantana, dwarf marigolds, nasturtium, petunia, salvia, sweet alyssum, ageratum, verbena, and trailing vines.

For shade use achimenes, tuberous and wax begonias, impatiens, fuchsia, and torenia.

For colored foliage use coleus and caladiums; for a trailer, German ivy.

WINTERFAT (*Ceratoides [Eurotia] lanata*).   A bit of an oddity, this shrub grows three feet tall and is covered with woolly hairs, white but becoming rust-colored with age. In fall the twigs are covered with woolly white fruits, resembling lamb's tails, and are wonderful for dried arrangements. (Plants of Southwest)

WINTER FLOWERS.   Winter gardens call for tough plants so don't think in terms of man-made garden hybrids. Alpines, such as crocus, are the obvious choice. In their natural mountain habitat there are two seasons, winter and August. This means they must bloom, ripen seed, and store food between the melting of snowbanks in late July and the new snows of September. Not an hour of this short summer can be wasted in preparation.

In our less orderly climate where winter temperatures go up and down, alpine bulbs still obey their mountain timetable. If you arrange for early thawing by choosing a naturally warm site you can often have "spring in January" by planting the pastel-shaded snow crocuses. There are many varieties of these and the colors range from white, yellow, and orange through blue and lavender.

The Korolkowii variety of snow crocus is one of the earliest blooming of all temperate climate flowers. This brilliant orange crocus often blooms before the winter's snow has melted, frequently blooming during a prolonged January thaw. It may even disappear beneath a late winter snow and suddenly reappear after the first sunny day.

WIRE WORM. The wire worm is repelled by white mustard, buckwheat, and woad.

WISTERIA *(Wisteria)*. This truly magnificent vine increases in beauty with every passing year. Drought-tolerant, it does well planted in a sunny location. *W. floribunda*, Ivory Tower, has long racemes of pure white flowers; when in bloom it resembles a waterfall.

Acetone extract from seeds of wisteria is somewhat toxic to codling moth larvae.

WITCH HAZEL *(Hammelia)*. This very fragrant, hardy ornamental blooms at a time when few other shrubs are blossoming outdoors. Its bright yellow flowers are not injured even in zero temperatures. On a cold, frosty morning witch hazels "shoot" their seeds as the seed pods crack open with a snap.

The well-known medicinal lotion is derived from an extract of the plant dissolved in alcohol.

WOOD FERN *(Dryopteris felix-mas)* SHIELD FERN. The powdered rhizome is toxic to armyworms.

WOOD ROSE *(Ipomoea tuberosa)* CEYLON MORNING GLORIES. The "wood roses" are the dried flower calyx (seed cases) of a yellow flowered, perennial, morning glory native to the West Indies and other tropical regions. The five so-called petals are usually the sepals rolled back from the central budlike ball. The wood rose looks like a rose carved out of wood, stiff and polished to a beautiful satin brown. These "roses" make spectacular corsages and are lovely in dried arrangements. (Parks Flower Book)

WORMSEED *(Chenopodium ambrosides)* JERUSALEM TEA. Some parts are toxic as extracts or dusts on several species of leaf-eating larvae.

WORMWOOD *(Artemisia absinthum)*. This name is loosely applied to many artemisias, but it properly belongs to "A.a.," a hardy perennial with woolly gray leaves and a strongly bitter odor. As a tea, spray it on the ground in fall and spring to discourage slugs and on fruit trees and other plants to repel aphids.

Southernwood is a close cousin, growing about three feet tall with gray green divided foliage. Silky wormwood *(A. frigida)* is excellent against snails. Plant it among your flowers in full sun and a dry location. Other family members include tarragon, mugwort, silver mound, fringed wormwood, and dusty miller. These are somewhat more moderate in odor than wormwood and southernwood.

WRINKLES. One of the most popular beauty preparations was invented a century before the birth of Christ by Galen, a Greek physician. Here it is: Melt four ounces of white wax in one pound of rose oil. Stir in a little cold water very carefully to give it a clear whiteness. Wash this mixture in rose water, and add small quantities of rose water and rose vinegar to make it the right consistency.

Rose water may be purchased at herb or specialty food shops. But if you want to make your own you can easily do so.

ROSE WATER

1 teaspoon rose extract   12 tablespoons distilled water

Measure carefully and use only distilled water. Mix liquids thoroughly and bottle, storing in a cool, dark place.

ROSE VINEGAR

1 pint white vinegar   1 cup fragrant rose petals
pinch of rosemary or lavender

Boil the vinegar and pour it over the rose petals. Add the rosemary or lavender. Cover tightly, let stand for ten days. Strain and pour into sterilized bottle.

Cook only in stainless steel, enamel, or glass pans and stir with a wooden spoon. Gather roses (wild or old-fashioned roses are best) when dew has dried but before the sun becomes warm. The green or white base of the petals, known as the heel, should be clipped off as this has a bitter taste. Press petals between two sheets of paper toweling to absorb moisture.

Glycerine and rose water, beloved of nineteenth-century ladies and responsible for many a beautiful skin well into old age, is again becoming popular; several of the leading cosmetic firms are now offering it. Other preparations such as Rose Milk are widely advertised. Here is a recipe:

ROSE HAND LOTION

Soak ¼ ounce of tragacanth (this is an herbal gum and may be obtained from Indiana Botanical Gardens listed as "Gum tragacanth") in water for four to five days. Mix 2 ounces of glycerine with 1 ounce of alcohol and add to the strained solution of tragacanth with ¼ to ½ ounce of rose water and 1 pint of water. If lotion needs thinning, add more water.

X

XERANTHEMUM (*Immortelle*). This is an interesting, papery, everlasting plant with double flowers in many colors. It grows to three feet and is lovely for flower arrangements, shadow boxes, or framed pictures.

Simply cut the flowers before they are fully open, remove leaves, and hang heads-down in dry location.

XOCHIMILCO. The Floating Gardens are the gayest and prettiest sight in Mexico City. The gardens are about three miles beyond the village of Xochimilco. (In fact, Xochimilco means "place where the flowers grow.")

The flowers have been growing there since the height of the Aztec Empire, when many nobles lived near Lake Xochimilco. Indians at the time built big rafts, covered them with earth, and planted vegetables and flowers. The rafts floated and the roots worked down through the earth into the water. These "floating gardens" gradually increased in size and became anchored by the interlacing roots of the plants. Now, the island rafts no longer float but are solid islands surrounded by canals. Flower vendors move about in the canals selling their colorful wares from canoes.

This type of garden was one of the earliest uses of what is now called hydroponics.

Y

YELLOW PUCCOON *(Hydrastis canadensis)* Goldenseal, Orange Root, Indian Turmeric. This is a perennial flowering plant of the eastern United States. It derives its name from its thick yellow root which is dried and ground for medicinal use. The plant has showy, lobed leaves sometimes as much as eight inches wide. Its inconspicuous greenish white flowers are ¼ inch across and are followed by red fruits in large clusters.

YLANG-YLANG *(Cananga odorata)* Flower of Flowers. The name comes from the pennant-shaped, long petalled flowers, and means "flower that flutters." The flowers are noted for their heady fragrance that permeates the air for a considerable distance around them.

The ylang-ylang is not a graceful tree, but its peculiarly shaped blossoms make up for any shortcoming in its stature. Just one of the fleshy, three-inch-long flowers will perfume a room.

As the flowers, almost hidden at first, reach maturity, the petals turn yellow. Then they gradually darken with age and their fragrance becomes proportionately stronger. The tree is generous with its flowers which, strangely, never fall; they just dry up and gradually blow away! The birds consider these delicate tidbits.

Plant this brittle, upright tree where it will be protected from strong winds. It is fast growing and will easily reach twenty-five feet within five years. Though listed as a rare tree, it isn't difficult to find in nurseries that specialize in unusual trees.

YOHIMBE *(Coryanthe yohimbe).*   Yohimbe is grown in West Africa. Its bark acts as an aphrodisiac directly on the nervous system, stimulating the spinal nerves and flow of blood. It also tranquilizes by reducing tension. Although yohimbe is not well known in the U.S., it is exceedingly well known in other parts of the world and highly valued for its quality and effectiveness as an aphrodisiac. Herbalists recommend that the yohimbe bark (of which a tea is made) be combined with vitamin C to produce yohimbine or yohimbiline ascorbate which is easier to assimilate. Unfortunately, the use of yohimbe is presently restricted by the FDA. Kava kava is a substitute.

YUCCA *(Yucca)* Amoles, Lily Family.   Yuccas are both beautiful and edible; the Indians even eat the flowers. The stalks are rich in sugar; the leaves produce a fiber used in making baskets and mats. The roots of the soaptree yucca *(Y. elata)* are used as a substitute for soap and shampoo. Eat the large, pulpy fruits of *Y. baccata* raw or roasted, or cooked and dried for future use. Cattle eat the flowers.

The yucca plant and the yucca moth are symbiotic. The word "symbiosis" means "living together." Any organisms that live together, whether they benefit one another, harm one another, or have no effect

The fibers of yucca leaves have long been used by Indians for making rope, matting, sandals, basketry, and coarse cloth.

at all, are symbiotic. The three forms of symbiosis are *parasitism* (as mistletoe), *commensalism* (a fungus or mold may live in the roots of a higher plant but do no harm), and *mutualism* where both parties benefit. In the instance of the yucca, the moth feeds but also pollinates, leaving enough seeds to start new plants.

The powdered leaves of spanish dagger (*Yucca shidigera*) are toxic to melonworms, bean leaf rollers, and celery leaf tiers.

Z

ZANTEDESCHIA *(Zantedeschia)* ARUM LILY, CALLA LILY.   This plant is easily grown in mild climates such as California where it sometimes "escapes" and grows wild. The most popular kind is the common calla which has handsome green leaves and bears large white flower spathes. It is cultivated in large quantities by market growers for the flower spathes, which are in great demand for decorative purposes, particularly at Easter.

The common calla and its varieties are moisture-loving plants and must be grown in rich, loamy soil that does not dry out quickly. In parts

Callas are such beautiful plants that it is difficult to believe that they grow wild in some areas. They're good both as houseplants and in the summer garden.

of North America where the winters are mild, the common calla and its varieties may be grown outdoors. They thrive in moist or wet soil on the edge of a pool or pond, or may even be planted in shallow water.

The chief kinds of yellow or golden calla are *Z. elliottiana* and *Z. angustiloba*. The pink calla, *Z. rehmannii*, is a smaller and much lower plant than the white or yellow flowered kinds. There is also a spotted calla, *Z. albo-maculata* and a black-throated calla, *Z. melanoleuca*. This last has yellow spathes with a conspicuous black purple spot at the base inside.

In Lapland, people grind the root of the marsh calla into flour for bread.

ZINNIA (*Zinnia*).   This is the easiest and most satisfactory annual to grow, and the hybridizers have made them so elegant that they can surely find a place in every garden.

Zinnias are bright and cheerful. The older varieties were beautiful in their day, but today zinnias are available in a rainbow of colors from white to purple. They even come in many striking bi-colors.

Lamb's-quarter gives added vigor to zinnias, as well as marigolds, peonies, and pansies.

The prairie zinnia (*Zinnia grandiflora*) is a spectacular bedding and border perennial shrublet. It forms ground-hugging cushions less than six inches tall and is completely covered with deep yellow flowers from midsummer through fall. Plant this close relative of the annual garden zinnias in full sun. Prairie zinnia is slightly toxic to celery leaf tiers.

ROSE HIPS FOR VITAMIN C

Many claims are made for vitamin C, a vitamin found in greater concentration in rose hips than in oranges. These claims have been made by authorities in the medical and nutritional fields and are certainly worthy of attention.

Most animals have the enzymes to synthesize their own vitamin C, but man and apes do not. This led Dr. Irving Stone and Dr. Linus Pauling to speculate that the proper dose for man was 2,000 mg. or more, and led to their recommendations of vitamin C in high doses to treat colds.

Dr. Fred Klenner of Reidsville, North Carolina, has accumulated vast clinical evidence that C is successful in treating a variety of viral and bacterial infectious diseases. C has also been long recognized as essential to the formation of the connective tissue called collagen. In this role it protects the gums from bleeding, the blood vessels from easy bruising, and it improves the healing of wounds.

Vitamin C also seems to neutralize some of the effects of cigarette smoking. One estimate is that one cigarette uses up twenty-five mg. of vitamin C. A heavy smoker who does not take supplements of vitamin C will be deficient in this vitamin.

In a study of 500 patients who had low back pain, Dr. James Greenwood, Jr. of Baylor University, found a significant number who were able to avoid surgery after taking approximately one gram of vitamin C per day. Their pain recurred when the vitamin C doses were stopped.

Dr. Atkins in his book *Super Energy Diet* writes: "In my own experience, vitamin C has served as a fatigue fighter, fatigue tending to reappear whenever a patient has neglected to take his usual dose. I also find it useful sometimes in the treatment of allergic conditions such as asthma and hay fever."

Gayelord Hauser in his book *Treasury of Secrets: How to Stay Young All Your Life*, states: "Some exciting studies have been made with vitamin C in connection with the serious condition of cataract, the clouding of the eye's crystalline lens. Several years ago Dr. Donald T. Atkinson of San Antonio, Texas, found that when he persuaded many of his patients who habitually lived on salt pork, corn meal, and coffee, to add fresh greens, oranges, and tomatoes to their diet, as well as eggs and other good proteins, the condition improved or the growth of the cataract was arrested."

These are just some of the benefits that doctors have found with the use of vitamin C so far. We are also told that, since the body does not store vitamin C, our supply must be constantly replenished; there is little danger of an overdose as the body eliminates what it does not use.

YOU CAN "GROW" VITAMIN C

Some rose hips, those of the *Rosa rugosa*, contain twenty times as much vitamin C as citrus fruit, and the wild Scandinavian types are even richer.

The rugosa blossoms with a single-petaled rose, a lovely flower both in bloom and in fruit. Planted eighteen inches apart, they make a bright living fence.

For either a hedge or a specimen planting, the method is much the same. If possible, set your roses out immediately after arrival (Farmer Seed & Nursery Co., Faribault, MN 55021, has an excellent selection). For individual plants, dig holes; for a fence, it's better to dig a trench about a foot across and one foot deep.

Put some well-rotted manure or compost several inches below where the roots will rest; this promotes a stronger start and quicker results, helping the roses produce usable hips much sooner.

Rugosas need little care but will establish more rapidly after transplanting if cut back. Leave three or four buds or leaf nodes on each stem. Rugosas, like other roses, respond well to mulching to retain moisture, especially during the summer. As it decomposes, the mulch also feeds the plants.

To receive the most benefit from this almost fantastic source of vitamin C, remember that the more roses you pick, the fewer the hips (these are the fruits which mature after the flower petals fall). Gather the hips when they are fully ripe, but not over ripe. If they are orange, it is too early. If dark red, it is too late. In the north, the hips usually ripen after they've been touched by the first frost.

After picking, cook your rose hips immediately and quickly to retain

In Europe, and increasingly in this country, people harvest their own supplies of vitamin C by collecting rose hips, which can be used in jam, soup, syrup, and marmalade, and made into rose hip tea.

the greatest amount of vitamin C. If this is not convenient, pack them in tight containers and keep refrigerated.

The hips, taken when fully ripe, can be split longitudinally and the inner seedlike structures removed. This gets rid of the hairs that are attached to them. The blossom end is usually removed and the pulp can be eaten raw or stewed, or can be used to make jam or jelly. Rose juice blended with apple juice makes a different but very tasty jelly. Be sure to cook your rose hips (and jelly) in glass or enamel saucepans.

Other sources for *Rosa rugosa* bushes are Stern's Nurseries, Inc., Geneva, NY 14456 (they also include growing information and recipes), Tillotson's Roses, Watsonville, CA 95076, and Kelly Bros., Dansville, NY 14437.

DRIED FLOWERS
BRIGHTEN ROOMS IN WINTER

Dried arrangements brighten a home or office during the cold months of fall and winter. During the fall gather weeds, vegetables, fruits, flowers, and foliage as well as dried grasses from the roadside, market, and garden. There is an abundance of material to select from. By treating these with certain special agents, you can preserve their beauty and color almost indefinitely.

Many plants will dry naturally as part of their life cycle. Tall brown stalks of mullein, teasel, thistle, heracleum (cow-parsnip), and non-poisonous sumac, cattails, and dock may be used without treatment if picked when dry. Dip cattails, however, in undiluted shellac to keep them from shattering.

Look for interesting pods on the ground beneath black locust. Sweet gum balls and cones of needled evergreens are attractive subjects. Also save some dried okra pods from the late garden.

Seed clusters of ornamental onions or leeks, as well as yarrow from the garden, dry naturally, but catch them before they start to deteriorate. Pick such materials a little prematurely; they can finish drying in the house. Hang them from coat hangers in your attic or in an unused room, or simply prop the stems in a deep, empty container.

Method for Hanging

Group three or four stems together and tie tightly. This is necessary because stems lose moisture as they dry, shrinking in size, and may loosen enough to slip out. Elastic bands or Twistems are better than string to keep materials bound. Hang the thick stalks such as mullein or large stems like cockscomb singly. Suspend the bunched materials or stalks upside down; this keeps the stems straight and the flower heads upright.

Attach in any manner that will allow free passage of air to all surfaces. Bunches or stalks may be hung on a line as you would clothes on wash day, or on a rack. You will need less space if you attach three or four bunches to a wire coat hanger. Clothespins are handy for attaching bunches to a line or hanger.

Choose any warm, dry spot for hanging provided it has free circulation of air. Use a kitchen, garage, attic, or shed if it is convenient. Don't shut the plants up in a closet or expose them to direct sunlight while they are drying.

Weather conditions at time of drying determine the number of days needed, but generally the majority of plants will dry in eight to ten days.

Plants that can be successfully dried indoors by hanging include acacia, artemisia, bird-of-paradise, rubber plant leaves, magnolia leaves, bells of Ireland, cacti, ornamental grasses, palm, pepper berries, nandina berries, snowball hydrangea, statice, strawflowers, and sunflowers (with petals removed). Globe thistle, liatris, and Queen Anne's lace can be dried by hanging but retain better color if buried in a drying agent. When hanging flowers, remove the foliage. Many dried materials will last much longer if sprayed a time or two with hair spray.

DRYING FLOWERS

Fragile garden flowers dry best buried in a special agent such as silica gel. This is sold under the name of "Flower-Dri," and comes in a kit (Park Seed Co., Inc.) complete with instructions, florist wire, and floral tape. To make your own drying agent, mix together equal parts of borax and yellow cornmeal, or use dry sand.

Pick your flowers on a dry day, selecting only those in good condition, preferably just before maturity. Process without delay.

1. Cut stems to within an inch of flower heads, and strip off any remaining foliage.
2. Make a false stem by dipping the end of a length of twenty-two-gauge wire in Elmer's glue. Insert this into the base of the flower head close to the stem and tape the two together.
3. Line a shallow box of suitable size with wax paper.
4. Prepare the drying mixture of borax and cornmeal. Add three tablespoons of uniodized salt per quart to mixture if you want better color preservation.

5. Cover bottom of lined box with about a ½-inch layer of mixture.
6. Place flowers face up in box, bending wire far enough below head so they lie flat. Mound up mixture beneath flower head to cushion.
7. Sift mixture gently between petals, adding gradually until flowers are just covered but not deeply buried. Do not crowd and do not dry more than a single layer of blossoms at a time.
8. Leave box uncovered. Check in six to seven days. Flowers should not be left in mixture when dry.
9. When drying is complete, gently brush away mixture. Then slide your hand carefully under each head and lift on your outstretched fingers. Place the flower on top of the mixture and allow it to remain for at least another twenty-four hours to firm up the petals. Without this treatment the petals will shatter. When completely cured, spray them lightly with a plastic coating for permanence.
10. Keep your dried flowers in closed boxes on tissue paper until you are ready to use them.
11. Store drying mixture in tightly covered tins to re-use.

Flowers to dry by burying include: acacia, China aster, bells of Ireland, celosia, daffodil, dahlia, delphinium, echinops, everlastings, gladiolus, liatris, lilac, lily, marigold, peony, Queen Anne's lace, snapdragon, stock, tulip, wandflower, zinnia. Roses may be dried if picked when only about two-thirds open; if full blown they will fall apart. Also try drying the buds of small "sweetheart" roses, they are especially nice for potpourri jars.

FOLIAGE

It is possible to preserve the lovely autumn foliage with glycerine. Leaves will achieve a rich brown shade but still remain soft and pliable. If the foliage has been in water before the processing with glycerine it may not be successful, so keep this in mind when making your selections.

Foliages which can be treated successfully include those of houseplants, as well as garden and florist materials. Here are some you may wish to try: acuba, aspidistra, beech, boxwood, cocculus (snailseed), copper beech, elaegnus, euonymus (*E. alatus* and *japonica*), ivy, laurel, leatherleaf viburnum, leucothe, magnolia (*M. grandiflora* and *virginiana*), mahonia, papyrus, peony, plum, red maple, sorrel tree, winter hazel.

There are no hard and fast rules as to which foliages can be preserved in glycerine, but it is best to use leaves or branches which absorb water freely and are entirely crisp and fresh.

Make a solution composed of one-third glycerine to two-thirds water.

Fill a container with this to a depth of four to five inches. Slash the stem of the leaf or branch with a knife or mash it with a hammer for about one inch at the cut end and absorption will be more complete. After placing stem or branch in the solution, allow it to remain until saturated. This is easy to determine as the color of the leaves will change. Store in a dark, cool place for about three weeks. When you notice beads of glycerine on the leaves, absorption is complete. Remove the branches and hang them upside down, allowing the solution within the stem to work down to the tip.

Adding a few drops of Clorox to the mixture will usually prevent mold from forming. And the solution is reusable. The foliage absorbs the solution more readily during the warmer months.

STRAWFLOWERS

It's so easy to have colorful cut flowers all the time with easy-to-dry strawflowers. Cut your material before the flower is fully open, remove leaves, and hang heads-down in an airy location. These are lovely in arrangements, shadow boxes, or framed pictures. Here are some suggestions:

Gomphocarpus fruticosus has bronze to greenish-yellow fruits on three-foot stems. Use this annual, which fruits in August, for both cut flowers and arrangements.

Statice, or sea lavender, is a perennial. Use fresh or dried for light, airy bouquets.

Acroclinum (sunrays) has rich shades of salmon, apricot, pink, rose, and cerise with white and creamy tones. Pick in bud to dry for winter bouquets, or use as fresh cut flowers. Plants branch and bloom freely and grow to twenty-four inches in height.

Echinops, or globe thistle, is a splendid everlasting. Globe thistles in shades of purple grow wild in many sections.

Xeranthemum has papery double flowers in many colors.

Helipterum sanfordii has silvery foliage and papery yellow flowers.

Rhodanthe has showy, nodding, 1½-inch flower heads of handsome rose pink.

Globe amaranth will thrive in any soil and stands drought well. It also retains form and color well.

Wood rose (*Ipomoea tuberosa*) is a must have! Dried flower calyx looks like a rose carved out of wood, stiff and polished to a beautiful satin brown. It blooms the second year when grown from seed.

NIGHT LIGHTS
AFFECT PLANT GROWTH

As if noise, air, and water pollution weren't enough, we are now discovering light pollution! Safety lights, installed by many people in their backyards, upset the timetables of plants and cause them to confuse night with day.

The amount of sunlight which a plant needs each day is called its photoperiod. Plants that flower when days are short are called "short-day plants," and include chrysanthemums, Christmas cactus, gardenia, kalanchoe, aster, and poinsettia. Plants that flower when days are longer are called "long-day" plants, such as marigold, petunia, black-eyed Susan, China aster, coneflower, feverfew, calceolaria, and weigela. And then there are the "day-neutral plants'" which have hormones to induce flowering whether days are long or short. In this category are African violets, roses, snapdragons, and tomatoes.

The growth pattern is altered when plants are near the night lights because the plants grow when they should be "sleeping." This is detrimental to trees, particularly in northern regions, because they continue to grow in the fall when daylight is shorter and they should be ceasing growth to prepare for winter. New growth is moist and tender. When a tree continues to grow well into the frost season, it becomes more sensitive and is more easily injured. Lighting also makes the leaves more sensitive to air pollution.

High-pressure sodium lamps are twice as efficient (for lighting purposes) as the mercury vapor lamp, and emit more red and yellow light. Mercury-vapor lamps, generally used on highways and city streets, give off a bluish green light which contains few red rays and many ultraviolet rays. Natural sunlight gives off light in the visible region from blue to green to yellow to red. The red region of the spectrum regulates the photoperiod.

Chlorophyll for photosynthesis — the food-making process that occurs only in nature and which is the chief function of the green leaves of plants — is activated by red and blue. The blue region also attracts night-flying insects.

Research by the United States Department of Agriculture indicates the red part of the spectrum is the growth-triggering light. During the twenty-four-hour period of a day, the light-dark cycles trigger the flowering, branching, dormancy, bulbing (as with onions), and other plant-growth responses.

In the Beltsville, Maryland, nursery, where tests were conducted, it was found that plants near the sodium lamps grew more rapidly into the fall season and also grew much later than plants of a like age that had been screened from night lighting. Trees which had been exposed to the light suffered severe winter die-back the following spring.

The sensitivity of seventeen species of trees to security lights (night lights), was rated by U.S.D.A. horticulturists. Trees tested having the highest sensitivity were Norway maple, paper birch, eastern catalpa, sycamore, American elm, and zelkova. The intermediate sensitivity group included red maple, gingko, honey locust, golden rain tree, Japanese pagoda tree, and littleleaf linden. Trees exhibiting low sensitivity were American holly, sweetgum, Austrian pine, Bradford pear, and willow oak.

These lights also have an adverse effect on gardens because plants need to sleep as well. Generally speaking, plants are affected within a radius of about twenty-five feet of the lights.

Laws affecting night lighting and lighting of commercial property vary from state to state but as our need for their protection seems to be growing, they are becoming more widely used. Therefore it is well to be aware of the effect they have on growing plants. With this knowledge you may be able to place your garden in a more favorable location or plant trees of varieties on which the effects may not be so disastrous.

GROWING WILDFLOWERS
FROM SEED

Wildflowers are truly wonderful, especially the mixtures. They are usually labeled by climate or geographical area and contain between six and twelve different kinds of flowers. Applewood Seed Company and Plants of the Southwest (see Sources of Supply) offer several such mixtures.

Wildflower seeds may be scattered on the ground but it is best to give nature an assist and rake them in lightly to give some protection from wind and rain. For better results till the soil and cover the tiny seeds with a thin layer of peat moss. Keep the seeds moist for about six weeks. On steep slopes where moisture is difficult to retain, sow the seeds into a top covering of very coarse gravel or lava rock. The seeds will sprout in between these materials which will help to keep the soil moist and to hold the small seedlings in place, thus giving their roots a chance to take firm hold.

TEMPERATURE

Most wildflowers germinate readily in a temperature range of 60°–75° F. Temperatures higher than this may be harmful to some species. This temperature sensitivity is nature's way of preventing seeds from germinating during hot, dry periods when it would be difficult for the seedlings to survive. Plant hardy perennials during spring or fall for best results. Seeds should be planted in a protected area to minimize the danger of being washed or blown away. Plant late enough in the fall to assure that germination will not take place until the following spring, or early enough so that seedlings are well established before the first frost.

Plant in spring after danger of frost has passed. Get the seed in the ground if you can before a rain, or water the seeds so that they have sufficient moisture to germinate.

DORMANCY

Seeds that fail to germinate under favorable conditions are said to be dormant. This state of dormancy is not accidental; plants survive in nature because of certain built-in timing mechanisms that delay germination until seedlings have the best chance for survival.

Some seeds will not germinate if exposed to cold temperatures. This causes a delay in germination, most usually until spring when rainfall and other conditions in the environment are favorable. Wildflowers having "cold temperature dormancy" may be planted outdoors in late fall, or be treated by a procedure called "moist-chilling" wherein seeds "overwinter" in your refrigerator for one to three months. Here is how this is accomplished:

Soak seeds that respond to moist-chilling in water at room temperature for twelve to twenty-four hours. Then mix them with a sterile, moistened medium such as sphagnum peat moss, vermiculite, or sand. Place in a plastic bag or similar container that is not airtight. Store them in the refrigerator at 40 ° to 50° F., not in the freezer, for three to six weeks. Keep medium moist but not wet. When chilling period is completed, sow seeds at once at relatively cool temperatures. Seeds that require only three to four weeks of moist chilling may be germinated in a greenhouse, if nighttime temperatures are in the range of 40°−50° F. Generally speaking, if this type of treatment is necessary, the packet will list the instructions. Most wildflower seeds do not need such exposure to cold temperatures and will sprout without special treatment.

MASS PLANTINGS

If you would like to make mass plantings of wildflowers, here is a simple way to achieve good results:

Till the soil to a depth of six to eight inches. The seedbed should have a loose, crumbly texture and good drainage. You can improve the air- and water-holding capacity of the soil by mixing in peat moss or other available organic material. Broadcast the seeds evenly and cover with a thin layer (not more than ¼-inch) of peat moss. Use a fine spray of water to moisten thoroughly. Keep evenly moistened for four to six weeks, thereafter waterings may be gradually reduced. If peat moss is not used as a top covering, rake the seeds lightly into the soil. If a few seeds are not entirely covered don't worry about them, it is best if the seeds are not deeply planted.

You may also sow seeds into a single layer of coarse gravel or lava rock (1- to 1½-inch size). This is a good way to plant steep slopes or areas difficult to keep moist. The seeds germinating in the cracks and crevices will be in contact with moist soil and also be protected from the elements. If natural rainfall is the only source of moisture, plant seeds in the spring just before anticipated periods of rain.

DRY AREAS

If you plan to plant a dry area, buy a mixture containing annuals, biennials, and perennials, most of which will sprout in ten to twenty-one days at a temperature of 55°–70° F. While such a mixture is best adapted to dry climates, most of the flowers will adapt to moist climates in sandy, well-drained soil. Perennials will survive cold winters in northern climates. Such a mix might contain: baby's breath, chicory, coneflower, cornflower, wild blue flax, gaillardia, penstemon, poor man's weatherglass, California poppy, prairie aster, and yarrow.

MOIST AREAS

Here is a mixture best suited for moist climates, but which will survive in dry climates if watered regularly. Recommended are baby blue eyes, columbine, coreopsis, dame's rocket, larkspur, ox-eye daisy, scarlet flax, and wallflower. These perennials will survive cold winter climates. In mountainous regions above 8,500 feet elevation, there is usually ample moisture for this mix.

HARVESTING SEED

To save seed of your wildflowers, break off stalks or seed heads, taking care not to disturb the root system. Timing is critical; if seeds are harvested too early, their viability may be seriously impaired. A change in color (often from green to brown or black) and a tendency to disperse seed are reliable indications of maturity. After removing your plant material dry thoroughly and either crush or shake to remove the seeds. Clean by sifting your seeds through a series of screens to remove dirt, chaff, and other unwanted material.

WILDFLOWERS ADAPTABLE TO LARGE AREAS

Most of these wildflowers are particularly well suited for restoration of large areas of land. They are easy to grow and, in most instances, highly adaptable to different climates and soils. These include: sweet alyssum (*Lubularia maritima*), prairie aster (tahoka daisy) (*Aster tanacetifolia*), baby blue-eyes (*Nemophila menziesi*), baby's-breath (*Gypsophila elegans*), black-eyed Susan (*Rudbeckia hirta*), catchfly (Campion) (*Silene armeria*), chicory (*Cichorium intybus*), columbine (*Aquilegia caerulea*), prairie coneflower (*Ratibida columnifera*), purple coneflower (*Echinacea purpurea*), lance-leaved coreopsis (*Coreopsis lanceolata*), plains coreopsos (*Coreopsis tinctoria*), cornflower (batchelor's button) (*Centaurea cyanus*), ox-eye daisy (marguerite) (*Chrysanthemum leucanthemum*), dame's rocket (*Hesperis matronalis*), wild blue flax (*Linum lewisii*), scarlet flax, (*Linum grandiflorum* var. *rubrum*), gaillardia (blanketflower, Indian blanket) (*Gaillardia aristata*), firewheel gaillardia (*Gaillardia pulchella*), gayfeather (blazing star) (*Liatris spicata*), gilia (standing cypress) (*Gilia rubra*), and rocket larkspur (*Delphinium ajacis*) (*Consolida ambigua*).

OLD ROSES OF
ROMANCE AND LEGEND

Once upon a time roses were different—less dramatically beautiful in form and color but far, far more fragant than they are today. Most of our roses are descended from these.

Fragrance is the rightful heritage of the rose. In the minds of most of us, fragrance and the rose can be inseparable. Even long ago when the rose was a simple flower, it was known as the Queen of Flowers. Surely it must have been the unsurpassed quality of its fragrance that gave it this prestige.

Recently flower lovers have been uneasy because of the scentless, or nearly so, roses appearing on the market. This trend toward mere beauty in roses is greatly deplored.

What is meant by the pure odor of roses, sometimes referred to as the "true old rose scent"? This is the property of that famous trinity: *Rosa centifolia*, the cabbage rose; *Rosa damascena*, the damask rose; *Rosa gallica*, the French rose.

This lovely scent has been inherited by many modern Hybrid Perpetuals and by some Hybrid Teas though in a lesser degree. The old H.P. General Jacquemont, which first saw the light in 1852, is the parent of a long line of deliciously scented roses. It is still popular with many because it possesses this true old rose scent.

Other available varieties which are endowed with it are: Hugh Dickson, Chateau de Clos Vougeot, Admiral Ward, Alfred Colomb, Duke of Wellington, Col. Oswald Fitzgerald Hadley, portadown, fragrance, Etoile de Hollande, C.K. Douglas, Arthur Cook, and flamingo.

The fragrance of a rose flower is in its petals. Many roses come in various tones of red and crimson. Red roses, perhaps because they are closer to the grand varieties of early times, are generally the most richly endowed with fragrance. Next come the pink varieties. Yellow roses are the least scented, and almost scentless.

GROWING OLD-FASHIONED ROSES

Old-fashioned roses are hard to find. Two excellent sources are Tillotson's Roses, Brown's Valley Road, Watsonville, CA 95076, and the Joseph J. Kern Rose Nursery, Box 33, Mentor, OH 44060.

Aside from the fragrance there is another advantage in growing old-fashioned roses. Many are not only hardy where winters are severe but also where summers are hot; and they are excellent for different landscaping effects.

Roses will grow, and grow well, practically anywhere if you are careful about a few things. Buy only first quality bushes, plant them with care in a sunny, well-prepared bed, maintain a regular dust or spray schedule, water and feed at correct intervals, and remove spent blossoms.

Pruning

For the pruning and care of old-fashioned roses, it is not easy to give definite rules. Each old, rare, or unusual rose is an individual with different types and habits of growth.

Old, shrub, and species roses should not be pruned in the spring as with the Hybrid Teas, for if you do, you will remove the canes which would have produced their great spring flowering.

However, roses that bloom repeatedly should have weak growth removed and be trimmed to shape the plant. Pruning them is more a matter of shaping and thinning rather than cutting back. Removing spent flowers encourages the growth of new flowering stems.

Treat varieties with but one annual flowering like flowering shrubs. Leave them alone, and put away your pruning shears until after they bloom.

Some of the loveliest and most intriguing of the old-fashioned beauties have long canes that arch over naturally from their own weight. Others have canes that grow straight up, which will bloom only at the top unless pegged or pruned.

To achieve a bushy, many-branched plant, shorten the long canes by one-third after the plant blooms and shorten lateral canes by a few inches. If you desire, keep this up until late summer, then leave the plant alone until after it blooms in the spring.

IF YOU WOULD MAKE A ROSE JAR

The great rose for making potpourri is *Rosa damascena trigintipetala*, which has just one annual flowering. This is also the rose used for attar of roses. It is available in this country. Plant garlic or onions with your roses; they are not only protective but actually increase rose fragrance when grown nearby.

Gather damask rose petals when the roses are blooming abundantly. Pack them in a glass jar which has a tight cover. The addition of the tiny pink buds of De Meaux will make the final product prettier and more highly scented. Between every two-inch layer of petals sprinkle two teaspoons of salt. (Use common uniodized salt.) Add more layers of petals and salt each day until jar is full. Keep in a dark, dry cool place for one week. Then spread the petals on a paper towel and loosen them carefully.

Mix the following ingredients thoroughly and mix well through the petals in a large bowl: ½ ounce violet-scented talcum powder, 1 ounce orris root, ½ teaspoon mace, ½ teaspoon cinnamon, ½ teaspoon cloves, 4 drops of oil of rose geranium. Add the following very slowly: 20 drops of eucalyptus oil, 10 drops bergamot oil, 2 teaspoons alcohol. Repack the mixture in the jar, cover tightly, and set aside for two weeks to ripen. It will then be ready for distribution into rose jars, which make wonderful birthday or Christmas gifts.

While some of these ingredients are readily available at most supermarkets, you may have difficulty finding others. Indiana Botanic Gardens, Box 5, Hammond, IN 46325, is a possible source of supply.

Roses are the most fragrant in the sunniest, most protected spot in the garden. It is there that they develop their essential oils in the highest degree. Collect the flowers before the sun is high, on a dry day after two or three days of dry weather. Never use inferior, rain-soaked blossoms or those that have been open for a few days. Also fragrant oils will not be present in the petals of flowers that have been in the house for a week.

DYEING WITH NATURE'S COLORS

Why use natural dyes? First, because they are so beautiful, second because of the wonderful feeling it gives you to say, "I did it!"

Natural dyes can be used on many types of material, yarn, cloth, macrame objects, crochet work, tie-dye fabric—silk, wool, cotton, jute— sometimes even wood. However, you will have your best results with cloth or yarn made of natural materials such as wool or cotton. Experiment if you like with others but try a sample first.

Some natural dye colors will be fast, others less so. Colors can be repeated, but don't count on the results being exactly the same; sometimes even commercial dyers have problems. Quantity dyeing is possible if large enough quantities of natural dyestuffs are available and your container is large enough.

Do all your dyeing in enamel kettles. Aluminum, tin, or iron pots change colors. Also do not use the mordants (mentioned later) in pots that you are using for cooking because some of these are poisonous. Keep a separate pan for dyeing and keep mordants out of the reach of children. Actually many of the flower and vegetable dyes do very well without mordants, but the colors will not be as permanent and sometimes not as bright.

There are probably hundreds of flowers, both garden and wildflowers, which can be used in dyeing, such as coreopsis (C. auriculata and C. calliopsidea), dock root (Rumex), goldenrod (Solidago), goosefoot (Chenopodium), hedge-nettle, betony (Stachys), wild geranium (G. robertianum), red flowered orchid cactus (Epiphyllum), Pansy (Viola tricolor), pearly everlasting (Anaphalis margaritacea), and iris (Iris). There are also many fruits and vegetables from your garden which can be used.

Many vegetables that make good dye can be served at the table and the cooking water saved for use as the dye bath. Over-cook spinach just slightly and it becomes a fancy purée for the table and a lovely green dye for your wool. Purple cabbage cooked for one hour is still edible and the broth is left for dyeing. From the cabbage you will derive several shades of green, depending on size and quantity of cabbage used.

Oranges are a real treasure. Use the orange juice or the pulp, boil the peel for one hour and dye your yarn or cloth a bright orange!

Save the water you boil your beets or beet tops in. It will dye your fabrics pink or yellow-green. Beets are fugitive and the color may fade a little, but the material will still be attractive. For a lovely rose to lavender color try cranberries—and use the berries with sugar added for sauce or preserves. Any of the nut hulls make good dyestuffs after the nut meats are used in cakes or cookies—particularly walnuts.

Blackberries, huckleberries, blueberries, strawberries, and raspberries all produce delightfully strong colors. For brightest purple try the raspberries, the equivalent of two 10-ounce packages will dye about ½ pound of yarn. Frozen grape juice gives a lively purple and is easy to use. The bottled grape juice gives an even darker shade.

Of the spices cinnamon, turmeric, ginger, saffron, paprika, curry powder, and even mustard give vivid dyes ranging from yellows to reds. Experiment with oregano, chopped chives, and others for different colors and effects.

Instant tea or coffee is quick acting and gives interesting shades of tans or browns. Rose hip tea produces a rosy tan that is delightful. Boil coffee grounds in a cheesecloth bag with wool for a rich chocolate brown. Even chewing tobacco, pipe tobacco, and snuff may be used for dark browns. Look in the barbecue section for hickory chips. Soak these overnight for the same rich brown that walnuts or hickory nuts give—and the chips are cheaper! For an attractive yellow green try sunflower seeds. After boiling them, spread the seeds outdoors somewhere and let the birds have a feast.

The canned goods shelf holds possibilities too. Canned spinach, beats, or okra (or blueberries), including the liquid in the can, gives pretty colors when simmered half an hour. Strain out the vegetables for table use, drop your thoroughly wet wool into the dye pot and you will have enchanting color. Cherries for pie give a soft rose and need only thirty minutes for cooking. Soak a box of currants overnight and simmer for a pearl gray or lavender color.

If you achieve a color you don't like or it's too pale, overdye it in another dye bath. The color will probably come out even more interesting.

Mordants are a must if you plan to use your dyed material in an article that will be washed frequently. A mordant is a mineral salt that binds the color, making it sunfast and washfast. If mordanting seems a bit difficult, use a half-cup of white vinegar or lemon juice or one tablespoon of salt in the dye bath. The material should then be washed gently by hand rather than machine-washed.

Dyer's mordants may be purchased at a prescription drug store, a chemical supply house or, occasionally, at a health store. To use them, mix the prescribed amount of the mordant crystals in a jar with about a cup of water. Shake them well to dissolve the mordant and add to dye bath in solution. The easiest ones to find and use are:

Alum (*potassium aluminum*). This does not change the color of the dye but binds the color to the material. Use with cream of tartar (spice counter). The quantity for ½ pound of yarn is two teaspoons of alum and one teaspoon cream of tartar.

Chrome (*potassium bichromate*). This brings out greens and yellows. But cover your dye pot because chrome is light sensitive and light exposure weakens it. Mix one teaspoon of chrome and one teaspoon of cream of tartar for a ½ pound of yarn.

Copper sulfate (*blue vitrol*). Copper sulfate intensifies green dye. Use one teaspoon to ½ pound of wool.

Iron (*ferrous sulfate or rust.*) This "saddens" or darkens colors. The same effect may be had by boiling a handful of rusty nails in a cheesecloth bag.

Tin (*stannous chloride*). Tin is good for brightening colors. It may be added to an alum bath during last half of cooking process to pick up bright color. Rinse in soapsuds after dyeing to keep tin from hardening wool.

Experiment for surprises. Sometimes a plain household ammonia rinse will give a unique shade. A lovely, soft rose from cranberries changed to a bright chartreuse after an ammonia rinse! After mordanting, wash your cloth or wool in a washing machine.

Here are the basic steps for dyeing a ½ pound of material—for more just multiply:

1. If using skeins of wool, tie them loosely in two places so they will not tangle—ditto for dyeing thread.
2. Soak skeins or cloth in water for at least an hour to keep wool from streaking or blotching in dye bath. Also dye will absorb more evenly.
3. Put dye bath (your vegetable, flower, or fruit broth) into an enamel pan. Add sufficient water to make three quarts. Add material, simmer one hour or until desired color is achieved. Remember colors look darker when wet so allow for this. Keep material submerged by poking with a wooden dowel or spoon. Wool is light and tends to float. Do not stir yarn or it will tangle.

Note: If using a mordant, dissolve in a jar with a cup or so of water, shaking to mix well. Pour this into dye bath and stir well before adding yarn or cloth.

4. Remove skein when desired color is reached. Pinch a little with your finger to get some idea of the color when dry. Rinse in hot water about the temperature of the dye bath (a change of temperature may shrink or mat wool). Rinse until water is clear and hang in a shady place for drying. If you mordant with tin, rinse in soapsuds and then rinse out soap. Each rinse should be a little cooler than the one before. Gently squeeze out water, never wring. A commercial fabric softener at the end of rinsing helps to make wool soft and fluffy.

If you have dyed wool yarn, use it for knitting, weaving, crochet, or needlepoint.

As you "get into" dyeing you may wish to try for variegated effects. Try dyeing half a length of a skein by placing a stick across the dye pot and letting half the skein hang in a cranberry bath, the other half length in a grape juice bath. Let each simmer the required length of time. Remove and rinse. You will have lovely soft colors that blend well with each other. Other combinations give equally interesting results. Try two shades of green or brown, yellow or orange.

Dyeing with nature's colors is fascinating and satisfying, and the colors you create will be uniquely yours. With these beautiful yarns and fabrics you can go on to make clothing or decorative objects.

Here are some suggestions for using nature's colors. Most of these work best with wool. Where indicated by a "c" you may also dye cotton:

For a soft shade of rose,
use beets for dying.

Common Name	Scientific Name	Plant Part Used	Dye Color
Acacia	*Acacia sp.*	flowers	Yellow or grayed maize yellow to light golden brown
Acacia	*Acacia sp.*	pods	Moss green or tan
Althea Shrub or Rose of Sharon	*Hibiscus syriacus*		Medium to dark blue
Anemone, Blue	*Anemone sp.*		Teal blue or light green
Bottle Brush	*Callistemon sp.*	flowers	Tan to greenish beige
Brass Buttons	*Cotula coronopifolia*		Deep brassy gold
Butterfly Bush	*Buddleia davidii*	flowers	Orange gold or gold green or golden brown – wool or jute
Butterfly Bush	*Buddleia davidii*	leaves and stems	Olive green
Butterfly Bush, in iron pot	*Buddleia davidii*		Various greens or black
Cactus	*Opuntia robusta*	purple fruit, steeped	Magenta to rose
Camellias, Red, in iron pot	*Camellia sp.*		Medium gray to dark gray
Camomile	*Anthemis nobilis*		Various gold yellows, aromatic
Canterbury Bells, Purple	*Campanula medium*		Medium green or pale blue
Chrysanthemum, Maroon	*Chrysanthemum spp.*		Variations of gray turquoise w/c
Daffodils, Yellow	*Narcissus pseudo-narcissus*		Bright yellow or deep gold
Dahlia	*Dahlia pinnata*	seed heads	Bright orange
Daisy, Brownish, Gloriosa, Black-Eyed Susan	*Rudbeckia sp.*		Bright olive green to dark green
Dock	*Rumex spp.*	blossoms	Rose beige to terra-cotta w/c
Dock	*Rumex spp.*	root in iron pot or with nails	Dark green to brown or dark gray w/c
Dodder and bits of Pickleweed Elderberry, Blue or Black, one pot and mordant method	*Cuscuta sp.* and *S alicomia sp.* *Sambucus spp.*		Yellow or ocher Mauve
Eucalyptus, Blue Gum	*Eucalyptus globulus*	leaves	Deep camel tan
Eucalyptus, Blue Gum	*Eucalyptus globulus*	leaves in iron pot	Light to dark green or charcoal gray w/c
Fennel	*Foeniculum vulgare*		Mustard yellow or golden brown, aromatic

Common Name	Scientific Name	Plant Part Used	Dye Color
Flax, New Zealand	*Phormium tenax*	flowers	Brown
Flax, New Zealand	*Phormium tenax*	seed pods	Bright terracotta
Foxglove, Purple	*Digitalis purpurea*		Chartreuse
Geranium, Red	*Pelargonium hortorum*	leaves	Dark purple to gray w/c
Goldenrod, in iron pot	*Solidago spp.*		Mustard color or tan orange or brown olive
Goosefoot, in unlined copper pot or with cupric sulfate	*Chenopodium sp.*		Dark green or green gold
Grape, Concord-type	*Vitis labruscana*	skins and ferrous sulphate	Dark blue
Hawthorne	*Crategus sp.*	blossoms	Variations of yellow green or variations of gold brown
Hedge-Nettle, Betony	*Stachys sp.*		Chartreuse green
Herb Robert, Wild Geranium, Red Robin	*Geranium robertianum*		Light golden brown to rich brown
Hibiscus, Red, and tin	*Hibiscus spp.*		Purple
Hollyhock, Red, in iron pot	*Althaeae rosea*		Brown
Hollyhock, Red, and tin crystals	*Althaeae rosea*		Wine color
Honey Bush, and tin crystals	*Melianthus major*		Violet
Indigo, Blue Pot	*Indigofera tinctoria*		Blue
Iris, Dark-Purple	*Iris spp.*		Various violet blues
Iris, Purple, Fleur-de-Lis, and tin crystals	*Iris germanica* and other *Iris spp.*		Various dark to light blues
Klamath Weed, and ammonia	*Hypericum perforatum*		Mustard gold or raw sienna
Knotweed, Doorweed, Mat-grass	*Polygonum aviculare*		Creamy yellow or brighter yellow or brassy yellow
Ladies' Purse, Yellow	*Calceolaria angustifolia*	flowers	Maize yellow to gold or deep orange, wool and jute
Laurel, California, Bay Tree	*Umbellularia californica*		Greenish beige, aromatic
Lichen, Brown Rock, Oyster Lichen	*Umbilicaria sp.*		Magenta violet, aromatic
Lilac, Purple	*Syringa spp.*		Light green or light blue green
Lobelia, Blue, in copper pot	*Lobelia erinus*		Pastel green
Lupine, Purple	*Lupinus spp.*		Bright yellow green or dulled green

Common name	Scientific name	Part/method	Color result
Manzanita	*Arctostaphylos spp.*	leaves	Deep camel or rose buff w/c
Marguerites, Yellow, Paris Daisy	*Chrysanthemum frutescens*		Gold or mustard green
Marigolds, with tin crystals	*Tagetes sp.*		Yellow orange or gold or dull green
Meadow Rue	*Thalictrum polycarpon*		Bright yellow, fragrant
Milkweed, Showy, and cupric sulfate or unlined copper pot	*Asclepias speciosa*		Moss green or brass green
Morning-Glory, Bindweed	*Convolvulus arvensis*		Dull green or khaki green to yellow
Mulberry, Black, tree	*Morus nigra*	berries	Intense red violet to dark purple or purple–wool and jute
Mule Ears	*Wyethia augustifolia*		Gold to brass
Mullein, and ammonia	*Verbascum thapsus*		Bright yellow or chartreuse
Nicotiana, Maroon, and cupric sulfate	*Nicotiana sp.*		Grayed green
Nightshade	*Solanum sp.*		Bright yellow or dull gold or various khaki greens
Oleander, Dark Pink	*Nerium oleander*		Light gray green or medium gray
Olives, raw	*Olea europaea*		Variations of maroon
Onion, Red	*Allium sp.*	skins	Gold to henna red to maroon w/c
Onion, Yellow	*Allium sp.*	skins, in iron pot	Yellow brown
Osage Orange	*Maclura pomifera*		Intense greenish yellow or deep burnt orange
Owl's Clover	*Orthocarpus spp.*		Lemon yellow or mustard or ocher
Pansy, Dark-Blue, steeped	*Viola tricolor*		Blue greens
Penstemon, Red, one pot and mordant method	*Penstemon sp.*		Medium brown
Petunias, Purple and English Walnut	*Petunia hybrida* and *Juglans regia*	leaves	Light khaki green
Petunias, Red, and Marigolds	*Petunia sp.* and *Tagetes sp.*		Various dark greens to brown
Pigweed	*Amaranthus sp.*		Moss green or brass or pale yellow
Pine Needles, in iron pot	*Pinus sp.*		Olive green, aromatic
Plum, Dark Red	*Prunus sp.*	leaves, and tin crystals	Violet to purple or lavender
Poinsettia	*Euphorbia pulcherrima*	leaves	Greenish brown
Primrose, Dark Red, in iron pot	*Primula sp.*		Greenish yellow or bright avocado

Common Name	Scientific Name	Plant Part Used	Dye Color
Rabbit Brush	*Chrysothamnus sp.*		Lemon yellow or gold copper
Ragwort, Tansy-Ragwort, Stinking Willie	*Senecio jacobaea*		Bright yellow or brassy gold
Redwood, California	*Sequoia spp.*	bark	Tan or light golden brown to terracotta
Rhododendron	*Rhododendron spp.*	leaves, in iron pot	Gray green
Rosemary	*Rosmarinus officinalis*		Various yellow greens
Rudbeckia	*Rudbeckia sp.*		Bright chartreuse to dark green
Sagebrush	*Artemisia tridentata*		Various tan golds or brilliant yellow or yellow
Salal	*Gaultheria shallon*	berries	Dark blue
Salal	*Gaultheria shallon*	berries, and cupric sulfate	Various dark greens
Santolina, Lavender Cotton, French Lavender	*Santolina chamaecyparissus*		Sienna gold or yellow
Scabiosa, Purplish, Pincushion Flower	*Scabiosa atropurpurea*		Bright green or dull dark blue
Self Heal, Heal-All	*Prunella vulgaris*		Bright olive green
Silk Oak	*Grevillea robusta*		Intense canary yellow or olive green
Snapdragon, Dark Reddish	*Antirrhinum majus*		Pale green or tannish gold
Spice Bush, and cupric sulfate	*Calycanthus occidentalis*		Light brown
Stock, Purple	*Matthiola incana*		Blue or turquoise
Tarweed	*Hemizonia luzulaefolia*		Golden yellow or light yellow; aromatic w/c
Tea, Black	*Thea sinensis*		Rose tan or gray or black
Tea, Sassafras	*Sassafras albidum*	bark	Light terracotta to orange tan
Twinberries, and tin crystals	*Lonicera involucrata*		Gray
Walnut, Black	*Juglans nigra*	leaves	Cinnamon to dark brown or tan to brown—wool, cotton, jute
Walnut, English	*Juglans regia*	catkins	Light golden brown
Woodruff, Sweet	*Asperula odorata*		Soft tan or gray green
Woolly Aster, Seaside	*Eriophyllum staechadifolium*		Bronze gold to golden brown
Yarrow	*Achillea millefolium and spp.*		Yellow to maize or dark green
Yarrow, in copper pot	*Achillea millefolium and spp.*		Chartreuse to tan greens

212

PLANT A BUTTERFLY BUSH!

The buddleia, or summer lilac, is called the butterfly bush because it attracts monarchs and swallowtails, including the tiger and zebra with their brilliant coloring.

Many plants that attract birds and bees also attract butterflies, but there are certain flowers, shrubs, and vines particularly loved by these "flying flowers." Such plants are attractive for their nectar.

Butterflies instinctively choose very different plants for egg-laying. These are mostly herbs, weeds, and certain trees; they include umbelliferous plants such as dill and parsley; weeds such as clover, goldenrod, milkweed, and dandelion; and trees such as willow, poplar, birch, and hackberry.

Butterflies, along with liking flowers whose nectar content is both ample and suitable, also have color preferences. They have a passion for yellow and purple blossoms, also notable in their likes and dislikes of weeds. Thistles are violet mauve, clover is mauve to rose purple, dandelions and goldenrod are yellow.

Butterflies don't care much for roses, particularly white ones, but will rush to the nearest purple lilac bush, preferring the purple to the white. They also like marigolds.

Gardeners who would attract butterflies should choose varieties which will encourage butterflies to feed, not breed. Among these garden attractants are wallflowers, alyssum, sweet William, sweet rocket, candytuft, mignonette, zinnias, and the beautifully scented phlox flowers. Or grow portulaca, butterflies love it.

As you will note from this selection, butterflies prefer the simpler blooms, some intensely perfumed, to the hybrids that have been bred away from their natural development. For this reason a butterfly-attracting garden is easy to grow.

In the strictest sense, butterflies aren't harmful. They cannot bite, chew, or sting. In the butterfly stage, they pollinate many flowers. Even in the caterpillar stage they are seldom numerous enough, in most areas, to do much damage. There is one in particular, however, which can be a real pest, the white or imported cabbage butterfly. This one is usually found wherever cabbages are grown (also other members of this family such as broccoli, collards, or Brussels sprouts). It is generally so abundant that its eggs, caterpillars, and chrysalides are readily discovered. This familiar butterfly is white with black dots on the wings and blackish front angles on the fore wings. They flit freely about over fields, meadows, and gardens, sipping the nectar of various early flowers through their long, coiled tongues. From time to time they light upon the leaf of a cabbage or other plant of the mustard family to deposit the small, pale yellow eggs which remain attached by a sort of glue.

About a week later the egg hatches into a tiny caterpillar which is very destructive. You will note its presence by the lacing of the leaves. A safe control for these worms is Bacillus Thuringiensis (B.T.) which is sold at garden centers under various trade names—Dipel, Biotrol, or Thuricide. Do not hesitate to use it. It kills only these caterpillars; otherwise it is harmless.

The worst enemies of butterflies are the flies and wasps which lay their eggs on the caterpillar or inside the body. When the eggs hatch, the larva eat the caterpillar. Other insects such as dragonflies and mantids, eat great numbers of butterflies and caterpillars. Spiders catch them in their webs or lie in wait inside flowers. Birds, frogs, toads, and lizards feed upon them.

Butterflies have no strong body parts to use as weapons against attack and are easily killed by their enemies. As a group they survive because of their high rate of reproduction. A female butterfly may lay several hundred eggs during her lifetime. Only a few will live to adulthood, but those few will carry on the species.

Butterflies are helped by protective coloration to escape from their enemies. The anglewings, for example, have bright colors on the upper surfaces of their wings but dull brown or gray underparts. When their wings are folded, only the dull underparts show.

Some butterflies, such as the monarch, feed on milkweeds, which makes them unpleasant tasting food for birds. And they advertise their bad taste by warning coloration. Other butterflies such as the viceroy, are protected from attack by having coloration similar to that of the monarch.

BEAUTIFUL EASTER EGGS, NATURALLY

Coloring eggs at Eastertime is a very old tradition practiced in many countries. The methods have varied, but none is more lovely or simple than the old German custom of employing natural materials. Perhaps best of all, in these days when we are again becoming ecology minded, these eggs may be eaten without worrying about their wholesomeness or their effect upon the system. And many of these materials are right in your own flower or vegetable garden, on your lawn, or in the woods or fields near your home.

Save the outer skins of onions. Carefully peel these off as they dry and darken. Store them in a mesh bag, using a bag of fairly close weave so small particles will not be lost. If you have a good supply of both yellow and red onions, save the skins separately for greater variation of color. Do not cook them together for the results will not be attractive.

Rainwater (or melted snow) makes the best onion skin broth. Catch this in a glass or enamel vessel and store it in advance. Well-washed glass or plastic vinegar jugs are handy for this purpose.

Make a broth by simmering the onion skins gently for an hour or until the color of the water is quite deep. Let cool to room temperature but do not remove the skins.

While the broth is cooling, remove the eggs you will use from the refrigerator so they, too, will be brought to room temperature; cold eggs may crack in the boiling process and spoil your efforts. Choose white eggs as large as possible. Some shells will not take up color as well as others, but this does not often happen.

Find an old sheet or several old pillow cases; those that are ready to be discarded are best as thin material absorbs well and permits the color of the skins to pass through to the egg. Tear into long strips, one inch wide and one yard long. If the material is still strong enough, tear the salvages into strips about ¼-inch wide. If not, have ready a spool of thin, soft white twine for use in tying the wider cloth securely after the eggs are wrapped. Or use sewing thread.

While the onion skins are cooking and the eggs are warming up, take your garden basket outdoors in the warm spring sunshine and see what you can find. Grape hyacinth makes a lovely delicate imprint; often the blue color is transferred to the egg as well.

Dandelion heads, carefully cut so they will lie as flat as possible, impart their own yellow color, and pink japonica or rose petals leave their own lovely hues. Hunt in the lawn or in a nearby field for young yarrow plants. Their fine, fernlike leaves leave an exquisite tracery. Clover leaves and ferns are lovely.

Do not overlook the decorative possibilities of dried grasses left over from winter and still holding their shape. Consider weeds with interesting outlines; some of these make markings on the eggs as pretty as cultivated flowers and plants.

Do not gather too much at a time. Some of your plants or flowers may wilt or curl before you can use them. You can always go back for more. Try, at first at least, for flowers and leaves that will lie as flat and close to the egg as possible. These leave a more definite imprint.

Now, lay a large soft bath towel on the table over which you will work. Have a pair of sharp scissors handy for cutting cloth or plants as you need them. Hold an egg in your left hand as shown in the sketch, slipping the cloth under the egg slightly so you can grasp it with your fingers. Lay a bit of fern, leaf, or flower on the cloth and fold it upward so that it is pressed securely against the egg.

As you lay each bit of flower or fern against the egg, pull the cloth over it and hold firmly. Then put on another flower or leaf, continuing in this manner until the egg is completely covered. Give a half-turn twist to the cloth when you have gone around the egg once (preferably lengthwise) so that you may also place a bit of plant on the ends. Do not be dismayed if you find this procedure awkward at first. With a little practice even children become adept. You may not be able to cover your first few eggs completely with flowers, nor is it even desirable to do so, for the contrasting brown is what makes the eggs so pretty.

After the egg is wrapped with the inch-wide strip of cloth, gently tighten the wrappings by going over them once more with the narrow cloth or twine. This is to prevent the covering from coming off in the coloring bath.

Use a slotted spoon to gently insert the eggs in the warm broth. Make sure they are well covered. For a four-quart saucepan three to five eggs at one time are all you should try to cook. After the eggs have been placed in the broth, cook them just as you would any hard-cooked eggs. Heat the broth slowly, letting it simmer for eight minutes so that the shells will take up as much color as possible.

When the time is up, take the eggs out one at a time with your slotted spoon. Flick any onion skins that may cling back into the saucepan. Place the eggs in water that is at room temperature or slightly warmer, and allow them to cool until they can be conveniently handled.

After a few minutes change the water to cool the eggs more rapidly or add more cool water. Or have two pans handy and just slip them into the other one as a new "batch" is made.

The wrappings should begin to slip off easily but, if necessary, untie them, pull off the cloth, and discard. (If you are pressed for materials, you can rinse them out in clear water, hang them up to dry, and re-use.)

After the wrappings have been removed, place the eggs on a couple of layers of absorbent paper to dry thoroughly. While they are still warm, add a bit of glamor by rubbing them with cooking oil, one that is not sticky. Use a soft cloth for rubbing. You will be delighted with the added shine which enhances the beauty of the coloring and the brown, or reddish-brown background.

These eggs, with their lovely, shadowy, imprints are perfect when used as centerpieces in low bowls or trays, or in a pretty, brightly colored basket.

If you wish, write names on these "flower eggs" with a wax pencil instead of, or in combination with, the flowers before cooking them in the broth. You may also cut out small pictures of thin (easily bendable) cardboard or heavy paper and wrap them on the eggs along with the grasses or leaves. Tiny bunnies and chicks are special favorites.

You need not have any qualms whatever about letting youngsters enjoy these eggs. Even if an egg should crack in the cooking process and a little color get on the egg, it is perfectly harmless.

The eggs can be made several days in advance of the time they are needed. After they have been cooked, return them to the egg carton and store them in the refrigerator. They will keep just as well as any other hard-cooked eggs. The oil film may dull a little when the eggs are cold but it will quickly become glossy again when they are taken out and returned to room temperature.

There are many other natural materials that can be used for coloring eggs. Some people use beet juice or even coffee grounds. In Russia the Pasque flower, *Anemone pulsatilla*, which imparts a green color, has been used to color Paschal or Easter eggs. In England, furze or gorse, a shrub with yellow flowers, has been used. Perhaps you will be inspired to experiment with other flowers and plants from your own garden.

NIGHT BLOOMERS
FOR DAYTIME WORKERS

For those who would love to enjoy their gardens but spend the daylight hours behind a desk, night bloomers are the solution. In the evening hours your garden is most hospitable and rewarding. And the odors of night bloomers are at their sweetest for this is one way they attract night-flying insects and ensure pollination.

For both beauty and fragrance in an evening garden, plant snow-on-the-mountain. Ghost plant is an excellent choice for the back border since it shoots up about two feet tall, grows easily from seeds, and needs little care.

The flower clusters of this plant are so dainty and tiny that you must look closely to see them. They grow in delicate rosettes on the branch tips and in the forks. But it is the white-margined leaves which surround them that are most showy and conspicuous on the upper stems of the plant.

In the southern and western states this plant often grows wild, covering whole hillsides and giving the effect of snow lying on the ground. They are extremely unusual to behold in the moonlight of a warm summer night. Old-timers in the West used the milky sap to brand cattle; it was said to be strong enough to remove the hair from the hide and yet not harm the animal.

Another choice for evening fragrance and beauty is roses. In front of the ghost plant (Euphorbia), plant McGredy's ivory, a dazzling white, and Sutter's gold. They are beautiful both day and night, and especially fragrant in the evening.

In front of your roses plant the low-growing border plant nicotiana. Choose white, which is pure white and tuberose scented. This lovely flowering tobacco is easily grown and blooms from June to August.

During the day the long-tubed white flowers hang down as though they were wilting but in the late afternoon they become large, milk-white stars and yield a most delicious fragrance.

In fact, most of the flowers that bloom at night are tubular. Therefore the nectar at the bases of the tubes can only be reached by those moths and insects which have very long tongues.

Another interesting fact is that the flowers of certain plants turn toward the moon at night, as the sunflower and heliotrope (*helios*, Greek for the sun) do during the day. When the moon is the orienting stimulus, it is called selenotropism (from *selene*, the moon); when a plant shoot turns toward the light, the effect is called phototropism.

One of the loveliest choices for a night bloomer is the moon vine (*Ipomoea noctiflora* Giant White, Burpee Seeds). This is an interesting, even entertaining, flower. The blossoms, which grow slowly during the day, open up in the evening with a quick twirling motion like pinwheels. As they spin open rapidly, they spill their delicious lemon-scented fragrance on the evening air to attract the gorgeous night-flying moths. This show takes place at approximately the same hour each summer evening. Once open, the great white flowers, several inches across, make a dazzling picture all up and down the vines.

If you plant moon vine, the beautiful sphinx moths which pollinate the flowers will arrive by the dozens.

Other flowering plants that bloom at night include evening primrose, the hardy thorn apple (Datura), sometimes called angel's trumpet, which stands up well in the heat of summer. Night-scented stock (*Mathiola bicornis*) is a drab looking plant by day but in evening the lilac-colored flowers expand and scent the air with a delicate perfume. At night many honeysuckles open wide their blooms, emitting their sweet scent to attract night-flying insects.

Night-blooming jasmine (*Cestrum nocturnum*), a tender shrub that grows five feet in height and bears very fragrant creamy-yellow flowers, is another favorite. Primarily a southern plant, it grows well in the north if given winter protection.

Marvel of Peru (*Mirabilis jalapa*), sometimes called four o'clock, opens only in cloudy weather or late in the afternoon; the funnel-shaped, white-, red-, and yellow-striped flowers close in the early morning.

Many of our most beautiful daylilies open in late afternoon and remain open at night, and they are almost always sweetly perfumed.

For those wanting the late-late show, plant the following:

Midnight to 2:30 a.m.: night blooming cereus in full glory.

3 a.m.: Amazon water lily is open.

4:30 a.m.: Virginia spiderwort is unfolding.

5 a.m.: purple morning glory opens, so does wild rose, Iceland poppy, and blue chicory.

5:20 a.m.: common blue flax is fully open.

TRY A TOUCH OF SILVER
IN YOUR GARDEN

Give some thought to the delightful contrast gray foliage can provide for your more brilliant plantings. The silver tones of these often overlooked plants add a luminous quality to a garden. On a hot day they add a note of coolness and restfulness. And their charm often becomes even more pronounced as they assume a frosty appearance in the moonlight.

Used with skill, gray-leaved plants provide a foil for vivid colors and complement soft pastels. For instance, the harshness of blazing bright red geraniums is made lacy, soft, and elegant in combination with the aristocratic dusty miller, Centaurea candidissima. These two plants grown together in full sun are almost breathtakingly dramatic.

And there are many members of this beautiful plant family. Cineraria diamond is a compact plant growing twelve inches tall with pure white foliage. As a border plant it is absolutely spectacular, has pretty yellow flowers in late summer, and looks elegant all by itself the rest of the time.

Another all-time favorite is silver queen, a special French strain with silver white leaves and very finely lacinated foliage. Its extra dwarf size makes it an excellent choice for window and porch boxes.

Then there is silver dust, also dwarf, whose fine, lacelike foliage provides a particularly lovely surprise when potted and grown in a carefully selected spot on porch or patio. Perennial candytuft snowflake, large flowered and pure white, provides cool contrast and a touch of class. The quiet, restrained beauty of these plants makes them synonymous with good taste, whether used alone or in combination. Pyrethrum silver lace (Chrysanthemum ptarmiciflorum) is one of the hardiest and finest textured of all.

For a plant that really glows try silvermound artemisia, a distinctive hardy plant with non-fading, fernlike foliage. This makes a perfect mound about 1 foot high and 1½ feet across. These plants sparkle and are wonderful for edging flower beds, adding contrast and exciting color to the garden. Combine with brilliant petunias, which also thrive in full sun, and you will have a breathtaking display.

Another member of the silver group not to be ignored is *Centaurea gymnocarpa*, whose finely divided, silvery white leaves have a woolly texture. Though it bears light lavender flowers during the late summer and early fall, this handsome two-foot-tall plant is grown principally for its foliage. A bit taller than some of the other family members, it is extremely attractive combined with cut flowers.

In rock gardens, dusty millers are the nearly perfect answer. Unusually charming with their low growing, closely packed mounds of molten silver leaves, the dwarfs all harmonize well with other rock garden favorites. The flower colors are enhanced and both intensified and softened by the contrast of their presence. *Cineraria maritima*, silverdust, with its finely cut silvery-white foliage and pygmy 9-inch height is particularly well adapted for this use.

Dusty millers are useful for blending together flowers because the surfaces of dusty millers' leaves are covered with long silky hairs which absorb some of the brilliant colors from other flowers. It is these silky hairs which also countribute to the gray effect, doing double duty as well in protecting the plants' surfaces and keeping them cool and moist in the same way that a mulch protects the soil. Furthermore, almost all gray-leaved plants withstand long periods of drought and intense heat and are rarely troubled by insects. Since many gray-leaved plants are also aromatic this probably helps to keep them pest free.

All the dusty millers need a sunny place and thrive best in light or well-drained soil. They prefer just a light touch of compost. And they are, themselves, a living mulch, shading the ground for nearby plants and actually protecting their roots.

Along with their other "sterling" qualities, dusty millers are easily raised from seed and may be sown directly out-of-doors in spring. Be careful to thin out the abundant seedlings sufficiently to allow adequate space for the plants to develop. This is about the only care they require. Seedlings will not bloom until the following year (once established dusty millers are considered perennials) but since these plants are grown primarily for their foliage—the flowers being somewhat inconspicuous—this does not greatly matter.

SUGGESTED READING

Ahrens, J. F. and P. M. Miller, *Marigolds: A Biological Control of Meadow Nematodes in Gardens*, Bulletin of the Connecticut Agricultural Experiment Station, No. 701, New Haven, CT, 1969.

Alther, Lisa, *Non Chemical Pest and Disease Control for the Home Orchard*, Garden Way Publishing Co., Charlotte, VT 05445, 1973.

Ascher, Amalie Adler, *The Complete Flower Arranger*, Simon & Schuster, New York, NY, 1974.

Baylis, Maggie, *Houseplants for the Purple Thumb*, 101 Productions, San Francisco, CA, 1973.

Best, Simon and Nick Kollerstrom, *Planting by the Moon*, Astro Computer Services, P.O. Box 16430, San Diego, CA. Distributed by Para Research, Whistlestop Mall, Rockport, MA 01966. Pub'd yearly.

Bicknell, Andrew, *Dr. Greenfinger's Rx for Healthy, Vigorous Houseplants*, Crown Publishers, New York, NY, 1980.

Bowness, Charles, *The Romany Way to Health*, Thorsons Publishers, Ltd., 37/38 Margaret St., London, W 1, 1970.

Brimer, John Burton, *Growing Herbs in Pots*, Simon & Schuster, New York, NY, 1976.

Clarkson, Rosetta, *Herbs, Their Culture and Uses*, The Macmillan Co., New York, NY, 1970.

Coker, Robert E., *Streams, Lakes & Ponds*, Harper Torchbooks, Harper & Row, New York, NY, 1954.

Coon, Nelson, *Gardening for Fragrance*, Hearthside Press, New York, NY, 1970.

– – – – – – –, *Using Plants for Healing*, Hearthside Press, New York, NY, 1963.

Davids, Richard C., *Garden Wizardry*, Crown Publishers, Inc., New York, NY, 1977.

Duruz, Selwyn, *Flowering Shrubs for Small Gardens*, Stephen Austin & Sons, Ltd., Great Britain, 1941.

Gordon, Jean, *Rose Recipes*, Red Rose Publications, Woodstock, VT, 1971.

Grae, Ida, *Nature's Colors*, Macmillan Pub. Co., New York, NY, 1974.

Harris, Ben Charles, *Eat the Weeds*, Barre Pub. Co., Barre, MA, 1975.

Healey, B. J., *A Gardener's Guide to Plant Names*, Charles Scribner's Sons, New York, NY, 1972.

Hunter, Beatrice Trum, *Gardening Without Poisons*, Houghton Mifflin Co., Boston, MA, 1964.

Jepson, Roger B., Jr., ed., *Organic Plant Protection*, Rodale Press, Inc., Emmaus, PA 18049.

Kraft, Ken and Pat, *The Best of American Gardening*, Walker & Co., New York, NY, 1975.

McWhirter, Norris and Ross, *Guinness Book of World Records*, Bantam Books, New York, NY, 1975.

Meyer, Clarence and David, *Fifty Years of the Herbalist Almanac*, Meyerbooks, Glenwood, IL 60425, 1977.

Millspaugh, Charles F., *American Medicinal Plants*, Dover Pub. Co., New York, NY, 1974.

Nehrling, Arno and Irene, *The Picture Book of Annuals*, Hearthside Press, New York, NY, 1966.

—————————, *Easy Gardening with Drought Resistant Plants*, Dover Publishers, Inc., New York, NY, 1968.

Newcomb, Duane, *The Apartment Farmer*, Hawthorn Books, New York, NY, 1976.

Novak, F. A., *The Pictorial Encyclopedia of Plants and Flowers*, Crown Publishers, New York, NY, 1974.

Ortho Books, Chevron Chemical Co., Ortho Division, 200 Bush St., San Francisco, CA 94104:
> *All About Ground Covers*
> *All About Landscaping*
> *Gardening With Color*
> *Houseplants Indoors/Outdoors*
> *How to Select and Care For Shrubs and Hedges*
> *The World of Cactus and Succulents*

Oster, Maggie, ed., *The Green Pages*, Ballantine Books, Inc., New York, NY, 1977.

Percival, Julia and Pixie Burger, *Household Ecology*, Prentice Hall, Englewood Cliffs, NJ, 1971.

Riotte, Louise, *Carrots Love Tomatoes*, and *Success With Small Food Gardens*, Garden Way Pub. Co., Charlotte, VT 05445, 1981.

———————, *Planetary Planting*, Astro Computer Services, P.O. Box 16450, San Diego, CA 92116.

Scully, Virginia, *A Treasury of American Indian Foods*, Bonanza Books, New York, NY, 1975.

Sunset Books, Lane Publishing Co., Menlo Park, CA, 1973:
> *Decorating with Indoor Plants*
> *Desert Gardening*
> *Gardening in Containers*
> *Greenhouse Gardening*
> *House Plants*
> *New Western Garden Book*

Tompkins, Peter and Christopher Bird, *The Secret Life of Plants*, Avon Books, New York, NY, 1972.

Wilder, Louise Beebe, *The Fragrant Garden*, Dover Publications, Inc., New York, NY, 1974.

Wilson, Helen Van Pelt, *Color for Your Winter Yard and Garden*, Charles Scribner's Sons, New York, NY, 1978.

Yearbook of Agriculture, 1972, U.S. Department of Agriculture, House Document 229.

SOURCES OF SUPPLY

Alpine Gardens
280 SE Firvilla Road
Dallas, OR 97338

Applewood Seed Company
333 Parfet St.
Lakewood, CO 80215

Armstrong Nurseries
P.O. Box 473
Ontario, CA 91764

Baldsiefen Nursery
Rhododendrons for the
  Connoisseur, Inc.
P.O. Box 88
Bellvale, NY 10912

Geo. W. Ball, Inc.
P.O. Box 335
West Chicago, IL 60185

Bedding Plants, Inc.
P.O. Box 286
Okemos, MI 48864

Bluestone Perrenials
Dept. 47
7247, OH 44057

Bountiful Ridge Nurseries
P.O. Box 250
Princess Anne, MD 21853

Breck's
6523 N. Galena Rd.
Peoria, IL 61601

Bunting's Nurseries, Inc.
Duke's Street
Selbyville, DE 19975

Burgess Seed & Plant Co.
905 Four Seasons Rd.
Bloomington, IL 61701

W. Atlee Burpee Seed Co.
Burpee Bldg. No. 12
Warminster, PA 18974

California Epi Center
P.O. Box 1431
Vista, CA 92083

DeJager Bulbs
188 Asburty St.
South Hamilton, MA 09182

Dutch Gardens, Inc.
P.O. Box 168
Montvale, NJ 07645

Dutch Mountain Nursery
7984 N. 48th St., R-1
Augusta, MI 49012

Emlong Nurseries, Inc.
2671 West Marquette Wds., Rd.
Stevensville, MI 49127

Farmer Seed & Nursery
818 N.W. 4th St.
Faribault, MN 55021

Ferris Nursery & Garden Center
811 4th St., N.E.
Hampton, IA 50441

Henry Field Seed & Nursery Co.
407 Sycamore St.
Shenandoah, IA 51602

Dean Foster Nurseries
R. #2
Hartford, MI 49057

Garden Way Mfg. Co.
102nd St. & 9th Ave.
Troy, NY 12180
(Tillers and Composters)

Goldsmith Seeds, Inc.
P.O. Box 1349
Gilroy, CA 95020

Grace's Gardens
10 Bay Street
Westport, CT 06880

Greenlife Gardens, Greenhouses
101 Country Line Rd.
Griffin, GA 30223

Gurney Seed & Nursery
Yankton, SD 57078

Hastings
434 Marietts St., N.W.
Atlanta, GA 30302

Hemlock Hill Herb Farm
Litchfield, CT 06759

Herbst Brothers Seedsmen
1000 N. Main St.
Brewster, NY 10509

High-Yield Gardening
401 Market Ave., N.
Canton, OH 44750

House of Wesley
2200 E. Oakland Ave.
Bloomington, IL 61701

J.L. Hudson Seedsman
World Seed Service
Box 1058
Redwood City, CA 94064

Indiana Botanic Gardens
Box 5
Hammond, IN 46325

Inter-State Nurseries, Inc.
504 E. St.
Hamburg, IA 51640

Jackson & Perkins
P.O. Box 1028
Medford, OR 97501

Johnny's Selected Seeds
N. Dixmont, ME 04932

J.W. Jung Seed Company
333 S. High St.
Randolph, WI 53956

Kelly Bros. Nurseries, Inc.
Maple St.
Dansville, NY 14437

Joseph J. Kern Rose Nursery
Box 33
Mentor, OH 44060

Kitazawa Seed Co.
356 W. Taylor St.
San Jose, CA 95110

Krider Nurseries, Inc.
P.O. Box 29
Middlebury, IN 46540

Lakeland Nurseries Sales
340 Poplar St.
Hanover, PA 17331

Henry Leuthardt Nurseries
Montauk Highway
East Moriches, NY 11940

Lewis Strawberry Nursery
Rocky Point, NC 28457

Lilypons Water Gardens
6885 Lilypons Rd.
Lilypons, MD 21717

Earl May Seed & Nursery Co.
208 No. Elm St.
Shenandoah, IA 51603

Mellingers Nursery, Inc.
West South Range Road
North Lima, OH 44452

Michigan Bulb Co.
1950 Waldorf NW
Grand Rapids, MI 49550

J.E. Miller Nurseries, Inc.
5061 W. Lake Rd.
Canandaigua, NY 14424

Musser Forests, Inc.
P.O. Box 340
Indiana, PA 15701

Nichols Garden Nursery
1190 N. Pacific Hwy.
Albany, OR 97321

Northrup King Co.
Consumer Products Division
Minneapolis, MN 55440

George W. Park Seed Co.
P.O. Box 31
Greenwood, SC 29640

Plants of the Southwest
1570 Pacheco St.
Santa Fe, NM 87501

Putney Nursery, Inc.
Box M.O.
Putney, VT 05346

Rayner Bros., Inc.
P.O. Box 1617-M
Salisbury, MD 21801

Roses of Yesterday and Today
802 Brown's Valley Rd.
Watsonville, CA 95076
(Catalog price $1.00)

Seed Savers Exchange,
  Kent Whealy
Rural Rt. 2
Princeton, MO 64673

R.H. Shumway Seedsman
628 Cedar St.
Rockford, IL 61101

Spring Hill Nurseries
6523 N. Galena Rd.
Peoria, IL 61601

Stark Bro's Nurseries
Box B2968A
Louisiana, MO 63353

Stern's Nurseries, Inc.
607 W. Washington St.
Geneva, NY 14456

Fred A. Stewart, Inc.
1212 E. Las Junas Drive
San Gabriel, CA 91778

Stokes Seeds, Inc.
737 Main St.
Buffalo, NY 14240

Sudbury Laboratory, Inc.
Sudbury, MA 01776

Thompson & Morgan
401 Kennedy Blvd.
Somerdale, NJ 08083

Tropexotic Growers, Inc.
708 60th St., N.W.
Brandenton, FL 33529

Tsang and Ma International
1306 Old County Road
Belmont, CA 94002

Twilley Seed Co.
Trevose, PA 19047

Van Bourgondien Bros.
245 Farmingdale Road
P.O. Box A
Babylon, NY 11702

Van Ness Water Gardens
2460 N. Euclid
Upland, CA 91786

Vermont Bean Seed Co.
Garden Lane
Bomoseen, VT 05732

Vernon, Lillian
510 South Fulton Ave.
Mount Vernon, NY 10550
(*Mortar & pestle*)

The Wayside Gardens Co.
Hodges, SC 29695

Western Maine Forest Nursery Co.
36 Elm St.
Fryeburg, ME 04037

White Flower Farm
Litchfield, CT 06759

Wilson Brothers Floral Co., Inc.
Box 192 M
Roachdale, IN 46172

Dave Wilson Nursery
Hughson, CA 95326

Wolfe Nursery
500 Termina Road
Ft. Worth, TX 76106

Worley's Nurseries
RD #1, Box 102 M
York Springs, PA 17372

INDEX

229